Nathaniel Wright Lord

Notes on metallurgical Analysis

Arranged for Students in metallurgical Laboratory of the Ohio State University

Nathaniel Wright Lord

Notes on metallurgical Analysis
Arranged for Students in metallurgical Laboratory of the Ohio State University

ISBN/EAN: 9783337185602

Printed in Europe, USA, Canada, Australia, Japan

Cover: Foto ©ninafisch / pixelio.de

More available books at **www.hansebooks.com**

✤ NOTES ✤

— ON —

Metallurgical Analysis

ARRANGED FOR STUDENTS IN THE METALLURGICAL LABORATORY
OF THE OHIO STATE UNIVERSITY.

BY

NATHANIEL W. LORD, E. M.

PROFESSOR OF MINING AND METALLURGY.

COLUMBUS, OHIO.
1893.

COPYRIGHT 1893, By NATHANIEL W. LORD.

PRESS OF HANN & ADAIR, COLUMBUS, O.

PREFACE.

THESE notes were written for the use of the writer's students in the metallurgical laboratory of the Ohio State University.

The object was to give in a condensed form the series of selected methods in metallurgical analysis which made up the course of study.

To the descriptions of the processes, such explanations have been added as experience has shown to be desirable for the assistance of the student in understanding the conditions necessary for accurate results.

Such methods only are given as have been tested by repeated use in the laboratory and found satisfactory.

No attempt is made to describe general reagents or apparatus, as students prepared to take this course are always familiar with all ordinary laboratory equipment and for special forms of apparatus reference is made to easily accessible books and papers.

The writer wishes to acknowledge his obligation to Blair, Troilius and other standard writers, as well as to numerous papers in the various technical and scientific journals, though it has been impossible to give credit in detail to all the sources from which material was taken in compiling these notes.

The references added are only those which it seemed important the student should consult for fuller information on the subject.

October 28, 1893.

CONTENTS.

	PAGE
Introduction	5
Obtaining and Preparing Samples for Analysis	7
The Analysis of Limestones	10
The Determination of Iron in Ores	15
The Determination of Phosphorus in Iron, Steel, and Iron Ores	20
The Determination of Silicon in Iron	35
The Determination of Manganese	38
The Determination of Sulphur	50
The Determination of Carbon in Pig Iron and Steel	59
The Determination of Titanium in Iron Ores	68
The Analysis of Coal and Coke	73
The Analysis of Furnace and Flue Gas	79
The Analysis of Blast Furnace Slag	83
The Analysis of Fire Clays	85
The Determination of Copper in Ores	88
The Assay of Ores for Zinc	91
The Analysis of Alloys of Lead, Antimony, Tin and Copper	93
The Examination of Water for Boiler Supply	96

APPENDIX.

Table of Atomic Weights	98
Table of Factor Weights	99
Applying Correction Factors of Volumetric Solutions to the amounts of Substance taken	99
Some additional Notes and Methods	100
1. Sampling Spiegel Iron and White Cast Iron	100
2. The Determination of Iron in Ores by Permanganate	100
3. On Difficulty in Filtering Solutions of Pig Iron and Steel in Phosphorus Determinations	100
4. The Purification of Barium Sulphate	101
5. On the Presence of Nitrites in Caustic Alkalies as a Source of Error in Sulphur and Carbon Determinations	101
Errata	102

INTRODUCTION.

BEFORE beginning the course in special analysis given in these notes, the student is supposed to be familiar with the ordinary qualitative reactions of the acids and bases, the preparation of reagents, and so much of the general methods of quantitative analysis as includes the use of the balance and weights, the ordinary operations of filtration, washing, drying, igniting and weighing of precipitates, the evaporation of solutions, and also the use and calibration of graduated glassware.

A careful study of the first two sections of Fresenius' System of Quantitative Analysis will be found of the utmost importance in regard to all these points of manipulation.

In addition to the above a few general precautions and explanations are necessary and should never be overlooked.

In adding reagents to produce any given effect it is important that the right amount be used. What this will be demands a thorough knowledge of what is to take place. In the descriptions of the various processes these amounts are approximately indicated, but it is impossible to provide in this way for all contingencies. Therefore, if the amount of reagent directed fails to do the work it must be increased or diminished as may appear necessary. Thus in every case where a precipitate is formed, it is *essential* that the filtrate be tested by a further addition of the reagent to make sure that the precipitation is complete. This is best done by adding the reagent to a small portion of the liquid in a test tube, and if a precipitate forms returning this to the main volume; often a little of the clear liquid over the precipitate can be tested in this way before filtration.

In all careful work the chemicals used should be tested as to purity. Many of the so-called "C. P." reagents are unreliable. A "blank" determination serves in many cases to detect this source of error. This is made by going through the process with the reagents alone, leaving out the substance to be tested. The amount of any impurity which would affect the result is thus determined and can be allowed for.

In testing a process it is always desirable, as was suggested by Dr. Wilson, to make duplicates, using different amounts of the substance. Agreement of results under these circumstances is obviously a much better guarantee of accuracy than when the same amount is used in each case.

The amounts of material prescribed in the description of the processes are those most generally used. They may be changed provided the reagents be varied to correspond.

A useful modification of many processes consists in the use of "factor weights." That is, weighing out an amount of the substance equal to some multiple of the percentage which the element to be determined makes of the precipitate weighed. Thus the actual weight will represent the percentage, and all calculation be avoided.

A table is added to these notes giving a series of such weights which may be substituted for those usually taken.

The same method may frequently be used to avoid the application of a factor of correction in the case of standard volumetric solutions.

NOTES ON METALLURGICAL ANALYSIS.

OBTAINING AND PREPARING SAMPLES FOR ANALYSIS.

The object sought by the technical analyst is to ascertain correctly the average composition of some certain lot of material — for example, a car load of ore.

The amount of material treated in the laboratory is of necessity limited to a few grammes.

The proper preparation of this small portion, that its analysis shall correctly represent the composition of the mass from which it is taken, constitutes the operation of "*sampling*."

The general mode of procedure is to take from the mass in question a large amount selected from different points, and containing coarse and fine material in the same proportion as they exist in the mass as a whole. This large sample which may weigh from 200 pounds to a ton, according to the amount of material the chemist has to examine, as well as to the *extent of variation permissible in results*, is then crushed to one-half inch or smaller, thoroughly mixed and subdivided by "quartering," until a sample of about ten pounds is obtained; this is pulverized and all put through a 6-mesh sieve, well mixed and again subdivided, till a sample of 100 or 200 grammes is obtained, which is put through a 90-mesh sieve and bottled for use.

Many variations will be necessary with different material. The following general principles may be stated:

1. As to size of original large sample. This must be greater as the material is less homogeneous and as the importance of the exact determination of any ingredient increases. Thus, a limestone can be

easily sampled; but a gold or silver ore consisting of small, detached fragments of a very valuable material in a valueless rock may require the fine crushing of the whole mass of ore and its careful mixing and subdivision, to secure an "average assay."

2. Materials of decidedly different specific gravities require great care to prevent separation into layers during mixing. "Quartering," that is, division of the flattened pile into quarters and then the separating of one of them, *all of which* must be carefully brushed together, constitutes a fair safeguard against this source of error.

3. Every particle of any portion must go through the sieve. The harder parts which are left unbroken toward the last, are of different composition from the softer and first pulverized portions, and if rejected would cause serious error.

4. Certain ores and slags contain particles of metal which cannot be pulverized. These must be kept by themselves, and the weight of the particles taken; also the weight of the portion of the sample passing through the sieve. The metal is then analyzed separately from the siftings, the two analyses combined in the ratio of the relative weights.

The sampling of metals presents many difficulties. Melted metals can be sampled during pouring by taking a little at the beginning, middle and end of the cast, and averaging the three analyses.

In general it may be stated—

1. Cast ingots are *not* homogeneous. Drilling from different portions will show different analyses. Hence, drillings from a number of points must be well mixed. A single "pig" of cast iron may vary largely from top to bottom.

2. In tapping a mass of metal from a furnace different portions of the "run" will show different compositions. Thus, a "bed" of pig iron will show wide variations in silicon and sulphur between the top and bottom of the cast.

3. In some metals the operation of drilling will result in a separation; for example, in drilling pig iron, the fine portion will be of different composition from the coarser; hence, careful mixing of the drillings is necessary.

"WEIGHING OUT" FROM THE LABORATORY SAMPLE.

In this operation the tendency of material of different specific gravities to separate must never be lost sight of. The substance should be carefully mixed together upon a sheet of glazed paper and small portions taken from different parts.

A second source of error is the separation of "coarse and fine," as in metal drillings. Great care is necessary to avoid serious difficulty here. The drillings may be moistened with alcohol so as to become adhereut and then small portions may be separated, to be subsequently (when dry) accurately weighed (Shimer).

The machinery for sampling may be quite elaborate.

References on methods of and machinery for sampling—

A. A. Blair — The chem. anal. of iron, 2d Ed., pp. 1-18; also p. 199.
Jour. an. and app. chem.. vol. V, p. 299 — sampling iron ores.
Jour. an. and app. chem., vol. II, p. 148 — as to irregularity of pig iron.
Eng. and Min. Jour., 1892, p. 75 — sampling machine.
P. W. Shimer, sampling cast iron borings, Trans. Inst. Min. Engs., vol. XIV, p. 760.
Wm. Glenn, sampling ores, Trans. Inst. Min. Engs., vol. XX, p. 155.
H. L. Benjamin, sampling machine, and also illustration of hand sampling, Trans. Inst. Min. Eng., vol. XX, p. 416.
E. K. Landis, iron ore sampling, Trans. Inst. Min. Engs., vol. XX, p. 611.

ESTIMATION OF MOISTURE.

Many materials as sampled in bulk are quite damp. Such (ores, clays, limestones, etc.) must be dried in a steam bath or by other means, and the loss of weight determined. This drying is best done on a portion of one or two pounds of the crushed and mixed material, which, after drying, is pulverized for the final sample. It is always well to determine moisture also in the final sample, and allow for it if present. The temperature for drying must not exceed 100°C.

The analysis may be stated on the " dry basis " and also calculated on the wet material.

For example: From a car load of iron ore portions were taken at different points, being careful to take proper amounts of lump and fine. The amount taken was 200 pounds. This was broken up as fine as beans, well mixed by shoveling and divided by quartering until a portion of two pounds was obtained, all being done rapidly to avoid loss of moisture.

This portion was weighed on an ore scales.

Weight........................925.4 grms.
After drying in a pan on a steam
  boiler — weight ....864.9 grms.

Loss .... 60.5 grms.

This was then pulverized, mixed and a portion of 100 grms. taken for the laboratory. This assayed—

Iron ....58.4 per cent.

Then 925.4 : 864.9 = 58.4 : X = the per cent. of iron in the ore in bulk.

It may be noted, 1st, that many ores will absorb water during pulverization. The water so absorbed will vary with the *weather;* 2d, complete drying of a large sample is very difficult; 3d, ordinary corked bottles are not moisture proof, and samples left in such will change in the course of time.

When much work is done special ovens for drying samples are of great assistance. For description of one used at the Edgar-Thomson Steel Works, see Blair chem. anal. of iron, p. 200.

THE ANALYSIS OF LIMESTONES.

The materials to be determined are "The insoluble silicious matter," consisting of sand (silica) and insoluble silicates, principally clay. Oxides of iron and alumina, usually not separated when small in amount, carbonate of lime and carbonate of magnesia. The iron is sometimes present as ferrous carbonate.

In examining limestone quarries to determine the quality of the stone for furnace flux, lime or cement manufacture, the rock should be sampled layer by layer as, different layers usually vary greatly from each other in composition, while material from the same layer (or "bed") is likely to be of moderately uniform composition. The stone generally ranks in quality according to the amount of carbonate of lime.

Process of Analysis.—Weigh 1 grm. of the finely ground sample. Transfer it to a 4-in. caserole or dish, cover with a watch glass, add 25 to 30cc of H_2O, then 15cc conc. HCl., warm until all effervescence has ceased, remove the cover, wash it off into the dish, add 4 or 5 drops of HNO_3 and evaporate the solution to dryness on a water bath or hot plate, or replace the cover and boil down directly over the lamp, using constant care to prevent loss by "spattering;" finally heat carefully over the lamp flame until all odor of HCl is gone and the $CaCl_2$ just begins to fuse; now cool, add 5cc of HCl, warm till the Fe salts are dissolved, add 50cc H_2O, heat until everything dissolves except the silicious matter, which forms a flocculent or sandy residue, filter through a 5 or 7 c. m. filter, wash thoroughly with hot water until a few drops of the washings show no reaction for Cl when tested with $AgNO_3$.

Ignite and weigh the residue, which, after deducting the weight of the filter ash, constitutes the "insoluble silicious matter," keep it for the determination of the silica by fusion.

To the filtrate, which should be about 100cc, add NH_4HO until it just smells distinctly of NH_3, should the precipitate be light colored and large in amount, add 5cc HCl and again NH_4HO as before, now boil the liquid until the smell of NH_3 has nearly gone — maintaining the volume of the liquid by adding water from time to time. Now remove the lamp, and let the precipitate settle, filter into a small filter and wash well with hot water, ignite and weigh the precipitate of, $Fe_2O_3 + Al_2O_3$.

The Fe_2O_3 may be determined in another portion and when deducted from the above will give the alumina by difference.

Dilute the filtrate to about 200cc, heat to boiling and add 40cc of a saturated solution of $(NH_4)_2C_2O_4$ first diluted with 40cc of water and heated to boiling also. Stir well, and let stand until the precipitate of CaC_2O_4 has completely settled, decant the liquid through a 9 c. m. filter without disturbing the precipitate, wash the precipitate once or twice by decantation, using about 100cc of boiling water each time, then transfer to it the filter and wash 6 or 7 times with hot water.

Dry the precipitate thoroughly, detach it as far as possible from the filter, put it in a weighed No. 0, porcelain crucible, burn the filter carefully on a platinum wire and add the ash to the contents of the crucible, now drop conc. H_2SO_4 on to the precipitate till it is well moistened, but avoid much excess. Heat the crucible (working under a "hood" to carry off the fumes) holding the burner in the hand and applying the flame cautiously until the swelling of the mass subsides and the excess of H_2SO_4 has been driven off as white fumes, finally heat to a cherry red for 5 minutes. Do not use the blast lamp! Cool and weigh the $CaSO_4$. The weight of the $CaSO_4$ multiplied by 0.735 gives the amount of $CaCO_3$ in the sample. The filtrate from the CaC_2O_4 should be, if necessary, concentrated by boiling to 300cc; should any MgC_2O_4 separate, dissolve it by adding a little HCl. Cool, add NH_4HO till alkaline, then add 10cc or a sufficient quantity of a saturated solution of Na_2HPO_4 then add gradually $\frac{1}{10}$ of the volume of the liquid of strong (26°) NH_4HO, stir hard for some time, cover and let settle until the liquid is perfectly clear (about 2 hours), filter and wash with water containing $\frac{1}{10}$ of its volume of strong NH_4HO and a little NH_4NO_3, dry, detach from the filter and burn the filter on a platinum wire; now ignite precipitate and filter ash in a porcelain crucible, using the blast lamp for 10 minutes.

The ignited precipitate is $Mg_2P_2O_7$ which multiplied by 0.756 gives the $MgCO_3$ in the sample.

Treatment of the Silicious Residue for the Determination of SiO_2.—Mix the ignited residue with 8 or 10 times its weight of dry Na_2CO_3, put the mixture in a platinum crucible of 15cc or more capacity, heat it over a Bunsen burner until the mass has caked together well, then over a blast lamp until it is in quiet fusion, now remove the crucible with a pair of tongs and dip the bottom in cold water, which will frequently cause the mass to loosen.

Wash off any of the material spattered on the cover of the crucible into a caserole with hot water, add the fused cake, if it has come loose; if not, fill the crucible with water and warm until the fused mass softens up and can be transferred to the caserole, finally clean the crucible with hot water and add the washings, if any material adheres so as not to be removed by washing, dissolve it with a little HCl and add to the rest, (on no account punch or "dig" the crucible out). When the fusion has been thoroughly disintegrated by the hot water and no hard lumps are left, cover the dish and add HCl until everything dissolves, warm till effervescence ceases, wash off and remove the cover and evaporate the solution to dryness on a water bath or otherwise, when dry and every trace of odor of HCl has gone, add 10cc dilute (1 : 1) HCl and then 50cc of water, warm till the Na Cl has dissolved, filter, wash well with hot water, dry and ignite the residual SiO_2. The ignition must be repeated and the residue reweighed until its weight does not change.

In the filtrate the iron, alumina, lime and magnesia may be determined as in the regular process and the amounts so found added to the weight of the main precipitates.

NOTES ON THE ABOVE PROCESS.

1. Limestones sometimes contain silicates which are decomposed by treatment with HCl, part of the SiO_2 going into solution as hydrate. This is made insoluble by evaporation to dryness. The presence of $CaCl_2$ renders the dehydration of the SiO_2 easy at the temperature of the water bath, a much higher temperature is to be avoided as silica may recombine with the bases and so either be redissolved on treatment with acid, or hold bases insoluble.

2. When the amount of residue is considerable it is often necessary to determine the silica it contains. In this case it is rendered soluble by fusion with excess of Na_2CO_3—the addition of $NaNO_3$ is to be avoided as it becomes caustic and attacks platinum ware. The fused mass, consisting of sodium silicates and aluminates, should be thoroughly soaked in water till it becomes soft and disintegrated, most of the sodium silicate being dissolved, then on adding HCl a clear solution without residue is at once obtained, if the HCl is added to the fused mass before disintegration it causes a gelatinous film of SiO_2 hydrate to enclose the pieces and arrests further action. The clear dilute solution on evaporation to dryness deposits the silica in a form more easily washed. A single evaporation of the SiO_2 solution even if dried for hours fails to completely render the SiO_2 insoluble, 97 to 98% only being recovered. The SiO_2 is not entirely pure, however, and these errors tend to balance each other. The impurity in the SiO_2 is largely alumina and the dissolved SiO_2 is largely precipitated with the alumina in the subsequent analysis.

3. The SiO_2 is hard to wash, retaining alkaline salts tenaciously it must be thoroughly washed with hot water till the filtrates no longer show a trace of Cl to $Ag NO_3$.

4. The SiO_2 must be ignited to *constant weight*, as it retains water most tenaciously. A blast lamp is necessary to remove the last traces.

5. Fe_2O_3 and Al_2O_3 are absolutely insoluble in solutions of NH_4Cl, but large excess of NH_4HO holds Al_2O_3 in solution to a small extent.

This is entirely separated by boiling off this excess of NH_4HO or by the presence of a large excess of NH_4Cl.

6. CaH_2O_2 is not completely soluble in NH_4HO unless sufficient NH_4Cl be present, it will be thrown down as a white precipitate, easily mistaken for alumina, so unless the Fe_2O_3—Al_2O_3 precipitate is small in amount it is well to redissolve it in HCl and reprecipitate by NH_4HO. If *much* Fe and Al are present the precipitate will certainly contain CaO, which must be removed by re-solution and reprecipitation after decantation and partial washing. Solutions containing NH_4HO and CaO will absorb CO_2 on standing, hence the $Fe_2O_3Al_2O_3$ must be filtered and washed promptly. On long standing a crystalline precipitate of $CaCO_3$ will sometimes be formed in the beaker with the $Fe_2O_3Al_2O_3$; of course in this case re-solution and re-precipitation is necessary. Distilled water sometimes contains CO_2 which will cause a precipitation in the same way.

7. *Calcium Oxalate* — Is very insoluble, but it is a difficult precipitate to filter and wash if not formed exactly right. On gentle ignition below visible redness it is changed to carbonate, and as such may be weighed, slightly too high a temperature however expels some CO_2. The complete conversion to *oxide* requires a very high temperature for a long time. The action of conc. H_2SO_4 on calcium oxalate converts it to a $CaSO_4$— the action is not violent and the excess of H_2SO_4 provided it is moderate, can be driven off without danger of loss by spurting. $CaSO_4$ will stand the cherry red heat of a Bunsen burner without alteration, the higher heat of a blast lamp will cause it to lose SO_3.

8. Magnesia will precipitate as oxalate in concentrated solutions, hence when much is present a re-solution of the calcium oxalate after a partial washing may be necessary. This is rarely the case, however, if the calcium is precipitated in properly diluted solutions. In the analysis of Dolomites the calcium precipitate must always be redissolved.

9. On concentrating the filtrate from the CaC_4O_4 a crystalline precipitate of magnesium oxalate will sometimes separate, this can be redissolved in HCl and added to the solution. If it contains any calcium oxalate it will leave a milky solution clearing slowly.

10. As the liquid in which the MgO is precipitated contains all the added material of the analysis careless addition of reagents may give so strong a solution of ammonia and soda salts that the precipitation of the magnesium phosphate will be incomplete unless the solution be largely diluted with water and ammonia, from such a volume of liquid the Mg will not wholly come down. In such a case the filtrate should be evaporated to dryness with excess of HNO_3 (3cc for each grm. NH_4Cl) to remove NH_4HO salts (which are thus decomposed into nitrogen and water) and then any remaining magnesia precipitated.

See Cook's "Select Methods Chem. Analysis."

11. There must be enough ammonium oxalate solution added to convert all the Mg. as well as Ca to oxalates or calcium oxalate will not completely precipitate.

12. Sodium oxalate, being very sparingly soluble, will sometimes separate with the Mg precipitate, when the solution is concentrated and much Na_2HPo_4 has been added; in this case the precipitate, after partial washing, must be dissolved in HCl and reprecipitated by NH_4HO.

References—
SiO_2 separation Jour. Anal. and App. Chem. Vol. IV, P. 159.
Mg. Ca. separation Fresenius, Quant., ¿ 73; ¿ 74; ¿ 107; ¿ 104; ¿ 154.

DETERMINATION OF IRON IN ORES.

Solution.—Most iron ores give up practically all their iron to hydrochloric acid, provided the acid be strong and the ore sufficiently finely pulverized.

There is danger of loss of iron by volatilization of Fe_2Cl_6 if a concentrated solution of Fe_2Cl_4 is boiled; hence, too great concentration of the solution and too hard boiling must be avoided.

If the residue from the action of HCl is white after ignition, the iron may be considered all extracted unless the ore contains TiO_2, in which case an insoluble compound of iron, phosphorus and titanium will remain, which may not color the residue. The presence of much titanium causes a milky appearance in the solution when diluted, and also causes the insoluble matter to run through the filter when washed.

In case of unknown ores it is safer to fuse the insoluble portion, as in a limestone, and determine the iron in this solution separately.

Process.—Pulverize the ore in an agate mortar, take a very little at a time and rub it until all trace of grit has disappeared when tested between the teeth or on the back of the hand.

Weigh out one grm., put it into a dry No. 0 plain beaker, brushing off the watch glass carefully, add 25 to 30cc concentrated HCl, cover the beaker with a watch glass and set on a hot iron plate, digest at a temperature just short of boiling until all the iron is dissolved, and on shaking around the beaker the residue appears light and "flotant" and free from dark, heavy particles. This may take from fifteen minutes to one hour or more, according to the ore, dilute the solution two or three times the volume and filter through a 5 c. m. filter into a 100cc flask. Wash the residue on the filter till it is free from acid. By care in letting the wash water run completely through each time, and not more than half filling the filter, this can be accomplished with not to exceed 80cc of filtrate, dilute to the mark and divide into two equal portions of 50cc, then determine the iron in each. The results should agree almost exactly, and their sum is the iron in one grm. of the ore. Dry and ignite the insoluble portion, if it is not white, or if suspected of containing iron, fuse as in case of the insoluble residue in a limestone. The iron in the HCl solution is then determined and added to the main quantity.

VOLUMETRIC DETERMINATION OF IRON BY BICHROMATE OF POTASH, WITH REDUCTION OF FERRIC CHLORIDE BY STANNOUS CHLORIDE.

The process depends on the following reactions:

1. A strongly acid solution of Fe_2Cl_6, if boiling hot, is almost instantly reduced to $FeCl_2$ by a solution of $SnCl_2$, the end of the reaction being judged by the disappearance of the yellow color of the iron solution.

$$SnCl_2 + Fe_2Cl_6 = SnCl_4 + 2\ FeCl_2.$$

2. Any slight excess of tin solution can be removed by adding $HgCl_2$, which is reduced to Hg_2Cl_2, forming an inert, white precipitate without action on iron or bichromate, the $SnCl_2$ being converted into $SnCl_4$.

$$SnCl_2 + 2HgCl_2 = SnCl_4 + Hg_2Cl_2.$$

This reaction is satisfactory, *provided* too much $SnCl_2$ is not present. The $HgCl_2$ *is in large excess* and the solution *not too hot*, otherwise metallic mercury may be formed as a gray precipitate, which will act on the bichromate solution and cause false results.

$$SnCl_2 + HgCl_2 = SnCl_4 + Hg.$$

This reaction is at once detected by the *gray color* of the precipitate and vitiates the results.

3. When a solution of bichromate of Potash is added to a solution of $FeCl_2$ containing a considerable excess of HCl the $FeCl_2$ is instantly oxidized to Fe_2Cl_6 with a corresponding reduction of bichromate.

$$K_2Cr_2O_7 + 14\ HCl + 6\ FeCl_2 = 2\ KCl + Cr_2Cl_6 + 3\ Fe_2Cl_6 + 7\ H_2O.$$

In the absence of an excess of HCl a basis chromium salt may separate as a precipitate and vitiate the results.

4. Solutions containing $FeCl_2$ strike an intense *blue color* with potassium ferricyanide, while solutions containing Fe_2Cl_6 give a yellow brown color. The ferricyanide solution must be fresh as it is reduced on exposure to light or on standing and it must be *dilute* or its own color will interfere.

PREPARATION OF THE SOLUTIONS.

1. Bichromate of Potash solution. Heat a sufficient quantity of the chemically pure salt in a platinum crucible, applying the heat carefully and avoiding all contact of the flame with the contents of the crucible (which would cause reduction), until the material just fuses to a dark liquid, withdraw the lamp at once and let the crucible cool. During cooling the bichromate will, after solidifying, gradually crumble to powder.

Of this powder weigh out exactly 8.785 grms., dissolve it in 200-300cc of cold water, transfer to a litre flask and dilute to 1 Litre. Of this solution 1cc should correspond to exactly 0.01 grms. of iron.

To test the solution, dissolve 1.4 grm. of *pure* ammonium ferrous sulphate in 50cc of water containing 5cc of HCl, run in 19cc of the bichromate solution from a burette and then, after stirring, put a drop on a white porcelain plate, add a drop of the ferricyanide solution, which will strike a blue color if the iron is in excess. Now add the bichromate solution drop by drop, testing the liquid after each addition until instead of a blue color it strikes an orange yellow. The liquid in the burette should now read 20cc, if not repeat the test and if the two results agree, make a factor of correction which must be applied to all results. For example, if instead of 20cc 20.2 were taken, then $20.2 : 20. =$ any reading : x the true reading or $\frac{20}{20.2} \times$ any reading $=$ true reading or percentage of iron.

2. Stannous Chloride Solution. Dissolve protochloride of tin ("muriate of tin") in four times its weight of a mixture of three parts of water to one of HCl; add scraps of metallic tin, and boil till the solution is clear and colorless. (Keep the solution in a closed dropping bottle containing metallic tin.)

3. A saturated solution of Mercuric Chloride. Keep an excess of the salt in a bottle and fill up with water from time to time.

4. A one per cent. solution of Ferricyanide of Potassium. This must be made fresh when wanted.

The other solutions keep indefinitely. The $SnCl_2$ solution must, however, be kept from the air.

PROCESS FOR THE ASSAY.

Transfer one-half the iron solution to a porcelain dish, add 5cc HCl, heat to boiling, and drop in the tin solution slowly till the last drop makes the solution colorless. Remove the lamp and cool the liquid by setting the dish in cold water. When nearly cold add at once 15cc of the mercuric chloride solution, stirring the solution with a glass rod. Let it stand three or four minutes. A slight white precipitate should form; none at all, or a heavy *grayish* one renders the results doubtful.

Now run in the bicromate solution until a drop of the liquid tested on the porcelain plate with a drop of ferricyanide solution no longer shows a blue, but a yellow color.

The number of cc used multiplied by 2 gives the percentage of iron in the sample.

Repeat the process on the second half of the iron solution running in at once nearly the full amount of bichromate and then finishing drop by drop. If the two results nearly agree, average them.

This analysis can be completed in an hour.

In the case of mill cinder and other decomposable slags, add to the finely powdered slag 20cc *water* and stir up well to prevent the cinder "caking" on the bottom of the beaker, then add 20-30cc HCl and proceed as before.

In the case of materials not attacked by HCl, fusion with Na_2CO_3 must be resorted to to get the iron in a soluble form.

Titanium, when present, will not affect this process provided care be taken to fuse the residue and add its solution to the main filtrate. When zinc is used to reduce the solution the TiO_2 will vitiate the results.

Titration with permanganate of potash after reduction by metallic zinc, is used by many chemists; for description of this process see Blair's Chem. An. Iron, p. 200; also for Clemens Jones's apparatus for reducing iron solutions by filtration through powdered zinc, see Trans. Inst. Min. Engs., Vol. XVII, p. 411.

SILICIOUS MATTER AND SILICA IN IRON ORES.

Treat one grm. of the finely pulverized ore in a caserole or 4 in. porcelain dish with 25cc conc. HCl evaporate the solution to dryness and heat on an iron plate until the residue is dry and scaly, dissolve in 10cc conc. HCl by warming, dilute, filter on a small filter, wash dry, ignite and weigh as "silicious matter." This consists of free silica, silicates, principally of alumina, sometimes titanium oxide and iron sulphide and occasionally barium sulphate.

Mix the ignited "insoluble silicious matter" with 6-8 times its weight of dry Na_2CO_3; fuse in a platinum crucible and extract with water, acidify with HCl, evaporate to dryness and heat on a water bath until all smell of HCl has gone. Add water and HCl and evaporate again to dryness; again take up with HCl, add water, filter and wash with hot water, dry, ignite and weigh the SiO_2. The second evaporation is not necessary unless very exact results are required, and in this case it is necessary to determine the impurity in the SiO_2 by hydrofluoric acid, as follows:

To the weighed SiO_2 in the platinum crucible add 1-10 drops conc. H_2SO_4 (enough to moisten it); then add conc. pure HFl until the SiO_2 is completely dissolved. Evaporate this to dryness under a good hood, dry and ignite the residue, deduct the weight of this from the weight of the SiO_2 first found and the difference is pure SiO_2.

The silica in the above process is completely volatilized as gaseous silicon fluorid, while alumina, iron oxide, titanic oxide and barium sulphate remain unaltered.

Should the silica contain alkaline chlorides due to imperfect washing, these will be converted to sulphates, and so the residue will be too heavy. This error need only be feared if there is considerable residue.

DETERMINATION OF PHOSPHORUS IN IRON, STEEL, AND IRON ORES.

The following methods are in general use and depend upon the separation of the phosphorus from the various bases, as phosphododeca molybdate of ammonia.

This substance, the so-called "yellow precipitate," when dried at 130.Cy has uniformly the composition $12 MoO_3 PO_4 (NH_4)_3$. This formula requires 1.65 per cent. of phosphorus. The average of many most carefully conducted experiments has shown that the precipitate contains 1.63 per cent. phosphorus within very narrow limits if free from admixed molybdic acid or other impurities.

The precipitate is only obtained pure when formed under very exact conditions, and is easily affected by subsequent treatment, so that all methods depending upon the weighing of the "yellow precipitate" or its volumetric determination must be carried out rigorously according to the prescribed directions in every detail.

When a solution of molybdate of ammonia in nitric acid is added to an acid solution containing phosphoric acid. The whole of the phosphoric acid is precipitated as the yellow "phospho molybdate of ammonia," under the following conditions:

1. All the phosphorus must be present as tribasic ("ortho") phosphoric acid.

2. A decided excess of ammonium nitrate or sulphate must be present.

The precipitation is most rapid when the solution contains between five and ten per cent. of the salt.

3. A certain excess of free acid must be present — preferably nitric, but *not* necessarily so. This must amount to at least 25 molecules of acid for each molecule of P_2O_5 present, and must be increased when sulphates are present.

4. Too great an excess of free acid must not be present, as this causes decomposition and partial re-solution of the precipitate. This action becomes perceptible when over 80 molecules of acid are present to each molecule of P_2O_5.

This action of free acid is prevented by an excess of molybdic acid. This excess must be greater as the amount of free acids is greater.

5. The yellow precipitate is *insoluble* in the solution of molybdate of ammonia in nitric acid; in solutions of ammonium salts, if neutral or only very *slightly* acids, but if *strongly* acid they attack the precipitate, which is, however, reprecipitated by the addition of molybdic acid solution to the liquid. It is also practically insoluble in a solution of potassium nitrate when neutral and not too dilute (containing at least 2 per cent.). Solutions of salts of organic acids usually dissolve the precipitate to some extent. From these solutions nitric acid and am-

monium nitrate, in some cases, reprecipitate the material in others, e. g., tartaric acid, oxalic acid — probably not completely.

The mineral acids, HCl, HNO_3, H_2SO_4, all have a solvent action on the precipitate even in the presence of ammonium nitrate. HNO_3 has the least, H_2SO_4 the most.

Pure water decomposes the precipitate to a slight extent and becomes milky, causing the material to run through the filter.

6. Precipitation is much more rapid from hot than from cold solutions, but in time it is probably complete at any temperature. The precipitate from hot solution is more crystalline and dense; from cold, more granular and fine, and harder to filter and wash.

7. *Agitation* greatly accelerates precipitation in this as well as in all other chemical reactions.

8. The precipitate dried at ordinary temperatures to constant weight retains a little acid and water, which it looses when dried at 130°C. By washing the precipitate with a neutral solution of ammonium or potassium nitrate it can be freed from all this acid without drying.

9. SiO_2 in solution does not seem to interfere with the complete precipitation of phosphorus if other conditions are right, but a small trace of the SiO_2 usually comes down with the precipitate, especially if the solution is too concentrated, or too warm, and stands too long. If the solution is rather dilute, not too hot, and is filtered promptly, the yellow precipitate can be obtained in solutions containing considerable SiO_2 practically free from it.

10. "Organic matter" has usually been supposed to interfere with the precipitation of phosphorus, but it is probable that in many cases, noticeably in steel analysis, the bad results attributed to this cause were due to the fact that the phosphorus was not all oxidized to the tribasic form.

11. When arsenic acid is present in the solution with phosphorus, it will be precipitated at the same time in amounts increasing with the temperature. Only very small traces come down at temperatures not exceeding 25°C.

12. Molybdic acid may separate with the precipitate as a light crystalline deposit. This free MoO_3 is only soluble with difficultly in acids and cannot be washed out of the yellow precipitate. Its separation must always be guarded against when the yellow precipitate is to be weighed or titrated. It forms when the amount of molybdic acid present is too great, the solution too concentrated, or too dilute, too strongly acid or too nearly neutral. Too high a temperature precipitates it. The addition of strong HNO_3 to a solution of molybdic acid will sometimes produce it, or adding molybdic acid solution to concentrated nitric acid solutions of iron.

13. When the yellow precipitate is thrown down in a solution con-

taining much iron and not sufficient acid, basic iron salts are liable to accompany it, making it reddish in color.

14. The yellow precipitate, if pure, is easily and completely soluble in NH_4HO, if it contains iron the solution will be turbid—from this solution the phosphoric acid is completely precipitated by "magnesia mixture" as $MgNH_4PO_4$. If the yellow precipitate contains any SiO_2 this will also, in part at least, dissolve in the NH_4HO and separate with the magnesia precipitate, making it a little flocculent. By cautiously adding HCl to the NH_4HO solution until nearly neutral before adding the Mg mixture, and letting it stand for some time in a warm place the SiO_2 separates completely and may be filtered off, then the phosphorus precipitated in the filtrate. In precipitating with magnesia mixture, add the reagent drop by drop and stir the liquid constantly, so that the precipitate separates slowly and in a crystalline form, otherwise it will be impure, containing magnesia in excess and molybdic acid. The $Mg_2P_2O_7$ must be ignited thoroughly with access of air to drive off any trace of MoO_3 it may retain.

For properties of "yellow precipitate" and effects of impurities and associated substances, see

Hundeshagen, Zeitschrift An. Chem., vol. XXVIII, p. 141; also Chem. News, vol. LX, p, 169.
Drown—Trans. Inst. Min. Engrs., vol. XVIII, p. 90.
Shimer—Trans. Inst. Min. Engrs., vol. XVII, p. 100.
Hamilton—Jour. Soc. Chem. Industry, vol. X, p. 904; also Jour. Anal. and App. Chem. vol. VI, p. 572.
Babbitt—Jour. An. and App. Chem., vol. VI, p. 381.
Precipitation by magnesia—
Gooch—Am. Chem. Jour., vol. I, p. 391.

GRAVIMETRIC METHOD FOR PHOSPHORUS WITH FINAL PRECIPITATION AS MAGNESIUM PHOSPHATE.

This is free from the chances of error due to the accidental impurity of the yellow precipitate. It is gravimetric throughout, and the phosphoric acid is finally weighed in a form not subject to variation in composition. It is independent of the kind of material treated and the per cent. of phosphoric acid; hence, is a standard method to which final reference must be made in all important determinations.

Process for Iron Ores.—In the absence of more than traces of titanium and arsenic. Weigh 1 to 5 grms., according to the percentage of phosphorus, of the very finely pulverized ore, put into a 4-inch porcelain dish or caserole, add 1cc HNO_3, then conc. HCl, using 15cc,+10cc more for each grm. of ore taken (∴ for three grms. 45cc), cover with a watch glass and warm till all the iron appears to be in solution, boil down to dryness, keeping covered to avoid spattering, dry on a hot plate till the acid is expelled, add 30cc conc. HCl, cover and digest till all the iron is dissolved. Now boil down until the liquid does not exceed 10 or 15cc. If the dish is kept covered there need be no formation of dry salt on the sides. Add water till the volume is 40 or 50cc, washing off the cover and the sides of the dish, filter through a small filter into a No. 1 or 2 beaker, transfer the residue to the filter and wash until there is no acid taste to the washings.

Dry and ignite the *residue;* for ordinary ores it is practically free from phosphorus, and may be thrown away if light colored and not too large in amount. In special or doubtful cases, however, fuse it, like the insoluble residue from a limestone, and after filtering out the silica, add ammonia to the filtrate, heat to boiling and let the precipitate settle, decant off the clear liquid, dissolve the precipitate in a few cc of HNO_3, add 20cc of molybdic acid solution and warm. Should a "yellow precipitate" separate, it must be added to that obtained in the main solution.

The filtrate and washings from the insoluble matter of the ore should not exceed 150cc. To this add 10cc of conc. HNO_3, then NH_4HO, until a precipitate is formed which does not disappear on stirring, then 3cc of conc. HNO_3 which must redissolve the precipitate and give a clear amber-colored liquid, not at all *red* in tint. The solution will now be quite warm. Add at once from a pipette in a fine stream 50cc of "molybdic acid solution,"[1] stirring the liquid vigorously all the time, and continue this stirring for about three minutes. Let the solution stand in a warm place until it is clear and the precipitate has all settled (which should not

---

1. *Preparation of the Molybdic Acid Solution.*—Add to 100 grms. of molybdic acid, 300cc of water, and then 120cc of strong (26°) NH_4HO. This will dissolve the MoO_3, and the solution must smell distinctly of ammonia. If it does not, add more NH_4HO. Unless the solution is clear, filter it, then dilute to about 800cc. Now mix 500cc of conc. HNO_3 with enough water to make about 1200cc. Cool both solutions and mix by *pouring the solution of molybdic acid into the diluted HNO_3.* The volume should now be 2000cc. Let the mixture stand a day or two, or until any small precipitate settles and use the clear liquid. If the solution of molybdic acid in NH_4HO is not diluted sufficiently, or if the above directions are not followed as to mixing, the molybdic acid may separate from the solution. 40cc of this solution will precipitate about 0.04 grms. of phosphorus.

require to exceed one hour), remove a portion of the clear liquid with a pipette and test it by adding a little more molybdic acid solution and warming to see if all the P_2O_5 is down.

Filter the liquid through a 7 c. m. filter, transfer the precipitate to the filter and wash until free from iron, with a 5 per cent. solution of ammonium nitrate *very slightly acidified* with HNO_3. The washing must be thorough or difficulty will be experienced in redissolving the precipitate, as in this case phosphate of iron and alumina may form and cause the filter to clog up. When the precipitate is washed put the beaker in which the precipitation was made under the funnel and redissolve the precipitate on the filter with dilute NH_4HO. When dissolved and the liquid run through wash the filter with water three or four times, then with a little dilute HCl, and then again with water. The filtrate should now be clear and colorless. If it is cloudy or colored (by a little iron), add HCl until the liquid is acid (the yellow precipitate usually separates), then add four or five drops of a saturated solution of citric acid, then NH_4HO to make the liquid strongly alkaline. This will give a clear liquid, the citric acid holding the iron in solution.

Now add drop by drop a considerable excess of "magnesia mixture,"[1] stirring the liquid constantly. This excess must be estimated from the probable amount of phosphorus in the ore taken. Continue to stir the solution vigorously for four or five minutes, then add NH_4HO until the solution smells strongly of ammonia.

Let it stand until the precipitate of $Mg\ NH_4PO_4$ has settled completely (one or two hours). The precipitate should be *white and crystalline;* if red or flaky, the results will be inaccurate.

Filter on a small filter or better on a Gooch perforated

---

1. *Preparation of "Magnesia Mixture."*— Dissolve 22 grms. of dry calcined magnesia in just sufficient dilute HCl. When dissolved add more of the magnesia until some remains undissolved, now boil; all iron oxide, alumnia and phosphoric acid will be precipitated. Filter the solution, add 280 grms. of NH_4Cl, 800cc water and 200cc conc. NH_4HO (26°). When all dissolves, dilute to 2000cc. Let stand a day or two and decant or filter the solution from any precipitate. 10cc of this will precipitate about 0.07 grms. of phosphorus.

crucible. Wash with water containing $\frac{1}{10}$ its volume of conc. NH_4HO and a little NH_4NO_3, dry, ignite and weigh as $Mg_2P_2O_7$, containing 0.279 of phosphorus.

It is essential that the filtrate from the magnesia precipitate should give at once a strong test for Mg (when tested with a drop of a solution of sodium phosphate), as a considerable excess of magnesia is essential to completely precipitate the phosphorus.

When ores contain titanium in any amount the solution on dilution before filtration may become turbid and the residue run through the filter on washing. In this case the residue will retain *iron and phosphorus.*

Modification of the Process on Account of Titanium.—First clear the filtrate from the insoluble matter by warming with the addition of nitric acid, and before adding ammonia, if it does not become quite clear, no matter, proceed with it exactly as before. After the filter paper containing the yellow precipitate has been washed out with NH_4 HO and HCl it must not be thrown away, as it may retain phosphorus and titanic acid, burn it and add what is left to the *insoluble residue.* Now mix the insoluble residue and the burned filter above mentioned with eight times its weight of dry $Na_2 CO_3$ and fuse as for silica. Boil the fusion with water until thoroughly disintegrated. The phosphorus passes into solution as *phosphate*, while the TiO_2 remains insoluble as titanate. Filter the liquid from the insoluble matter, acidulate the filtrate with HNO_3, evaporate it to dryness, add a little HNO_3, then water and filter from the separated SiO_2.

Add to the filtrate 25cc molybdic acid solution and warm, filter off the yellow precipitate and treat it exactly like that from the main solution. Add the phosphorus thus obtained to that obtained from the first solution.

See a valuable paper by Drown and Shimer, Trans. Inst. Min. Engrs., vol. X., p. 137.

When ores contain arsenic there is always danger that the final results will be high from the presence of magnesium arsenate.

In this case proceed as follows: To the filtrate from the insoluble residue, which should be in a small Erlenmeyer flask, add a solution of Na_2CO_3 until the liquid becomes dark colored, then add small portions of pure crystallized sodium sulphite (Na_2SO_3), which must be free from phosphorus. (This is best done by dissolving the salt in water 1 to 5, add HCl until the solution reacts slightly *acid* and use this solution). Warm the solution, shaking occasionally until up to boiling. If any precipitate forms redissolve it by a few drops of HCl. By this time the solution should be *colorless* and all the iron reduced to the ferrous form; if not, continue the warming. Finally add 10cc HCl, and boil until all the odor of SO_2 has gone (usually about three minutes). Remove from the lamp and pass a stream of H_2S gas through the liquid for fifteen or twenty minutes, or till all the As_2S_3 is precipitated. (The volume of the liquid should not exceed 150cc.) The As and any Cu present separate completely as *sulphides*. Filter the solution rapidly into a beaker and wash with a little H_2S water. Now boil the filtrate till all the odor of H_2S has disappeared, then add HNO_3 drop by drop to the hot liquid until the change of color shows all iron to be changed to the ferric form, and the liquid becomes perfectly clear. A faint cloud of separated sulphur may form, but will disappear on heating and does no harm. From this point proceed exactly as with the filtrate from the insoluble residue in ores when As was not present.

The process depends upon the complete precipitation of arsenic by H_2S in hot strongly acid solutions. The reduction of the iron is necessary to prevent a large separation of sulphur from the action of the H_2S on the Fe_2Cl_6.

The As_2S_3 precipitate may be used for the determination of arsenic, provided the solution has not been boiled too long before precipitation by H_2S, which will cause volatilization of As_2Cl_6. See Fresenius, Quant. Anal., for details.

DETERMINATION OF PHOSPHORUS IN BLACK BAND ORES AND OTHERS WHICH CONTAIN MUCH CARBONACEOUS MATTER.

These should be weighed out in a porcelain crucible and carefully burned, taking care not to heat so rapidly as to cause loss by blowing out of fine particles. After all "flaming" has ceased, turn the crucible on its side and heat with access of air till the carbon is gone and an "ash" is left. Treat this by the regular process. Avoid a high temperature in burning or the material will *cake*, thus causing imperfect solution. A dull red heat is sufficient.

DETERMINATION OF PHOSPHORUS IN MILL CINDER.

Two points here need attention. *First*, the material being a soluble silicate, it should be decomposed by weak acid and evaporated to dryness, as in the case of the determination of silica after fusion.

Second, all mill cinder contains particles of metallic iron which contain phosphorus as phosphide, and would give off PH_3 gas when dissolved in HCl, so HNO_3 must be used to oxidize the phosphorus. Proceed as follows: Weigh 1 grm. into a porcelain dish, add 20cc HNO_3 1.2 sp. gr., stir well to prevent caking and warm till action ceases, then add 10cc H_2O and 10cc conc. HCl. Evaporate to dryness and heat on an iron plate to 200°C for half an hour. Add 10cc HCl, digest till all dissolves. Dilute, filter, and proceed as with an ore.

DETERMINATION OF PHOSPHORUS IN IRON AND STEEL BY THE MOLYBDATE-MAGNESIA PROCESS.

The phosphorus in iron and steel exists principally as *phosphide*. When these metals are treated with ordinary solvents, the oxidation of the phosphorus is incomplete; when treated with non-oxidizing acids (HCl H_2SO_4), part of the phosphorus passes off as gaseous H_3P. Even concentrated HNO_3 probably fails to convert all the phosphorus into tribasic phosphoric acid.

These metals also contain carbon compounds which pass into solution in HNO_3 as a dark colored substance, and the presence of this dissolved carbonaceous matter is generally supposed to interfere with the precipitation of the yellow precipitate, though it seems probable from recent experiments that, if the phosphoric acid were in the tribasic state, this organic matter would be without influence. It is certain, however, that unless the oxidizing action is strong enough to destroy completely this carbonaceous matter, the phosphorus will not be all oxidized and can not be all precipitated.

The most generally used and oldest method of oxidation is the "dry method." It consists in dissolving the metal in HNO_3 concentrated, or, more generally, dilute (1.2 sp. gr.) evaporating the solution to dryness. The dry mass of basic ferric nitrate is then heated to about 200°C for some time. At this temperature the salts are decomposed, the iron converted largely to ferric oxide, and the dissolved carbon and phosphorus completely oxidized. This residue can then be dissolved in HCl and treated like an ore. The method is certain and independent of delicate reaction.

Several "wet" methods of oxidation have been proposed and used with apparent success. Among these are oxidation in nitric acid solution by chromic acid, and by permanganate of potash. Aqua regia fails to cause complete oxidation, as probably does $KClO_3$ and HCl. When permanganate is used there is a separation of MnO_2 which holds phosphorus insoluble and must be all dissolved by adding a reducing agent, such as oxalic acid or ferrous sulphate. Wet methods are quicker, but must be restricted to the class of material to which their adaptability has been proved.

Process for Phosphorus in Iron and Steel.—Take from 1 to 5 grms. of the well mixed borings. Treat them in a covered caserole or dish with 25 to 75cc of HNO_3 1.2 sp. gr. (made by mixing water and conc. HNO_3 in equal parts). Add the acid cautiously to prevent boiling over, heat till action has ceased, boil down to dryness, using care to prevent spattering, and keeping the dish covered. When dry, set on a hot iron plate and heat the dish to about 200°C for from 30 minutes to 1 hour, at the end of this time the material should be hard and scaly and show no trace of acid fumes. Now add from 15 to 25cc conc. HCl, digest till all the iron is dissolved and proceed as in iron ores. Many *steels* will leave no residue insoluble in HCl. In this case filtration is unnecessary.

The phosphorus retained in the residue in case of irons and steels practically amounts to nothing.

To make the filtration easy, add to the above HCl solution about 50cc H_2O, boil this solution 5 minutes and then let settle completely, decant off the clear liquid, transfer and wash the residue with warm water adding a little HCl at first, this treatment seems to cause a consolidation of the SiO_2 residue.

DETERMINATION OF THE PHOSPHORUS BY WEIGHING THE YELLOW PRECIPITATE.

This method is very generally used as a "rapid method."

That the yellow precipitate obtained may be of constant composition, the details of the process must be carried out exactly as given.

As the yellow precipitate contains 1.63% of phosphorus it is convenient to take that quantity of the material in grammes — as in that case the weight of the precipitate in grammes will be the per cent. of phosphorus.

The drying and weighing of over 0.4 grm. of yellow precipitate is difficult, hence for ores having over $\frac{1}{5}$% of phosphorus take $\frac{1}{2}$ the above amount (0.815 grms.).

The following details are essentially those given by E. F. Wood in an important paper in The Zeitschrift fur Anal. Chem., vol. 25, p. 489:

Process for Iron Ores.—Weigh 1.63 grms. of the finely pulverized ore into a 4-inch dish or caserole, add 25cc conc. HCl, digest and evaporate dry, heat as in the first process.

Now add 20cc HCl, digest till all the iron is dissolved, add 30cc H_2O, boil, settle and filter into a No. 1 beaker, wash with small portions of water, letting each run through before adding the next, the volume of filtrate and washings need not exceed 70 or 80cc. Now add 35cc concentrated HNO_3, boil down rapidly until the volume of the liquid is 15cc (judged by putting this amount of water in a similar beaker and comparing the two), take off the hot plate, wash off the cover and add water, so that in all from 15 to 20cc of water are added, that is, at least as much water as solution, stir the solution and add from a pipette 40cc of molybdic acid solution which should be at a temperature of not less than 25°C so that the mixed solution shall be at a temperature of not less than 40°C. Stir vigorously for 2 or 3 minutes, set in a warm place (*not* on a hot plate or water bath, which may cause precipitation of molybdic acid) until the precipitate has settled and the liquid is perfectly clear — this will take 30 minutes to an hour according to circumstances.

Fold a small filter of about 4 c. m. diameter, put it in an air bath and dry at 110°C for 15 minutes — remove to the balance and weigh it rapidly to the nearest milligram, do not "swing" the balance but simply weigh at a standstill.

Place the paper in a small funnel, filter and transfer the precipitate to it. This may be done more rapidly if the clear liquid be first drawn off by a small siphon or, as can be done by a little practice, with a pipette, so as to leave but a few cc of liquid above the precipitate, if care be taken not to disturb the precipitate, there is no loss whatever, and the filtering of 30 or 40cc of liquid can be avoided. Wash the precipitate carefully with water containing 1% of HNO_3 (six times at least) set the funnel and contents in an air bath and dry at 110° to 120°C 30 minutes after all visible moisture has disappeared. Then remove to the balance and weigh rapidly as before. The difference between the second weight and the first gives the weight of the yellow precipitate, and this in grammes gives the phosphorus in per cents.

Process for Iron and Steel.—Weigh 1.63 grammes of the well mixed borings into a 4-inch covered dish or caserole, add cautiously 35cc HNO_3 sp. gr. 1.2 boil to dryness, then "bake" on a hot iron plate at 200°C for thirty minutes. Take up by heating with 20cc HCl, then proceed exactly as in the case of ores.

Many steels will dissolve without residue — of course in this case filtration of the HCl solution is unnecessary.

The above methods assume that the phosphorus all passes into solution and that arsenic is absent. Titaniferous and arsenical ores and metals for this reason must be treated by the first method, fortunately they are the exception.

Many devices have been proposed for shortening this standard method — such as employing wet methods of oxidation — to avoid baking and evaporation. Before adopting any one, it must be carefully tested against standard methods on the kind of metals the chemist has to work with, and only used so far as it is thus proved adaptable. Thus Mr. Wood, at Homestead, uses chromic acid, with the nitric acid for solution and does not bake. See Blair Chem. Anal. Iron and Steel, 2d edition, p. 102.

VOLUMETRIC METHODS.

Determination of phosphorus in iron ore, iron and steel by precipitation as Phospho molybdate, and determination of the amount of the precipitate by estimating the molybdic acid it contains volumetrically, either by potassium permanganate after reduction by zinc, or by neutralizing with standard alkali.

TITRATION BY PERMANGANATE — "EMMERTON'S METHOD."

When the yellow precipitate is dissolved in NH_4HO, and then mixed with a very considerable excess of H_2SO_4 it all remains in solution. If this solution be treated warm with metallic zinc, the zinc rapidly dissolves, hydrogen being given off and the "molybdic acid" MoO_3 is rapidly reduced, giving a dark red and finally olive green solution containing what would be if the reduction were complete, Mo_2O_3, but which appears to be in fact $Mo_{12}O_{19}$. Now, if to this solution, after being rapidly filtered from the undissolved zinc, a solution of permanganate of potassium be added, the $Mo_{12}O_{19}$ is instantly reoxidized.

The solution becomes colorless, and finally, when oxidation is complete, a drop of permanganate in excess gives a permanent pink tint.

While there is some uncertainty as to the nature of the oxide produced by reduction, this possibly being different for different workers, it is proved by the extensive use of the method that working always in exactly the same way, the *reduction is uniform*, and hence, the titration is a reliable method for estimating the yellow precipitate, and indirectly the amount of phosphorus.

Consult a paper by F. A. Emmerton, Trans. Am. Inst. Min. Engrs., vol. XV, p. 93.

Process of Analysis—Preparation of the Solution of Permanganate.—Assuming the reaction between the oxidizing agent and the reduced solution of the yellow precipitate to be $Mo_{12}O_{19} + O_{17}$ (from the permanganate) = 12 MoO_3, and that the yellow precipitate contains 24 MoO_3 to 1 of P_2O_5, as is now undisputed, there must be 17 atoms of oxygen furnished by the permanganate to oxidize the reduced molybdous acid equivalent to one atom of P in the yellow precipitate.

As it is difficult to weigh out permanganate and dissolve it without change, it is necessary to make up the solution of approximate value and then determine its strength. This is easily done by titrating against a solution containing iron.

Two atoms of iron, as FeO, require one atom of oxygen to change them to Fe_2O_3; hence, 34 atoms ($= 2 \times 17$); of iron as ferrous salt will consume as much oxygen from permanganate as will the reduced molybdic acid, equivalent to one atom of P in the yellow precipitate.

One atom of phosphorus = 31 — 34 atoms of iron = 1904 (56×34); hence, $31 : 1904 = 1 : 61.41$; that is, $\frac{1}{61.41}$ of the amount of iron to which the permanganate solution is equivalent will give the amount of phosphorus to which it is equivalent when titrating yellow precipitate as above.

One molecule of $K_2Mn_2O_8$ furnishes five atoms of free oxygen on reduction; hence, to furnish 17 atoms of oxygen as above $3\frac{2}{5}$ molecules of permanganate are required which will be the amount equivalent to one atom of phosphorus in the above process.

The molecule of $K_2Mn_2O_8 = 316$, $3\frac{2}{5}$ times this equals 1,074, the amount equivalent to one atom = 31 of phosphorus; or 34.6 parts by weight of permanganate will be equal to one of phosphorus to be determined.

Therefore, to make a solution of permanganate, of which 1cc shall be equivalent to .0001 gram. of phosphorus, or to 0.01 per cent. when one grm. of substance is taken; dissolve 3.46 grms. of the salt in water and dilute to one litre.

Allow the solution to stand some time before using, as its strength is liable to change at first, but gradually becomes nearly constant. Finally determine its value exactly, first against pure ammonium ferrous sulphate, and second, against a sample of steel in which the amount of phosphorus has been exactly and repeatedly determined by the magnesia process.

To standardize against the iron salt dissolve 0.8597 grms. of the $Fe(NH_4)_2(SO_4)_2$ $6 H_2O$ in 200cc of water containing a little H_2SO_4; add the permanganate solution from a burette until the last drop gives a permanent pink tint and then take the reading.

This amount of the iron salt contains 0.12282 grms. of iron, which will reduce as much permanganate as would be equivalent to .002 grams. phosphorus; therefore, 20cc of the solution should be required. If more or less is taken, calculate the amount of phosphorus to which 1cc is equivalent by the proportion $20 : n = .0001 : x$, n being the amount used and x the value sought.

The determination of the value against a known steel is desirable, as it gives a result which is independent of all assumptions as to the nature of the oxide produced by the zinc reduction.

Treatment of the Samples of Iron or Steel.—Weigh five grms. of steel or one to five grms. of iron, according to the amount of phosphorus; put it into a 4 in. dish or caserole, add 25 to 75cc of HNO_3 1, 2 sp. gr. Add the acid cautiously to avoid boiling over. After action has ceased cover well and boil down dry, then bake on a hot plate 30 minutes, then add 20 to 40cc conc. HCl. Heat till all the iron oxide is dissolved, then boil down to 15cc, being careful to avoid the formation of any dry crusts on the side. This is accomplished by keeping the dish well covered and shaking it around in the hand a little.

Now add 20 to 40cc conc. HNO_3, washing off the cover with it into the dish. Boil down again to 10 or 15cc. It is essential that no dry iron salt form on the sides. This is easily avoided by covering with an inverted watch glass a little smaller than the dish, so that the condensed acid will flow down the sides and keep them clean. Cool slightly, moving the liquid around so as to dissolve any crusts of ferric nitrate formed. Now add 30 to 50cc of water and filter into a 400cc Erlenmeyer flask. The volume should be about 75–100cc. Steels do not require filtration as a rule.

Now add NH_4HO until the ferric hydrate separates and the mass becomes thick and smells of ammonia. Then add gradually strong HNO_4, until the precipitate redissolves and the liquid has a clear, *amber* color, not the least red. The volume should now be about 250cc, if not, dilute to that amount, then put a thermometer in the liquid and raise the temperature to 85°C. Now add at once 40cc of molybdic acid solution. Close the flask with a rubber stopper, wrap it in a thick cloth and shake violently for five minutes.

This violent agitation, combined with the high temperature, causes the yellow precipitate to separate at once in a particularly dense and easily filtered form.

At the end of this time let settle an instant, then uncork the flask and filter off the solution, using a 9 c. m. filter. Wash flask and precipitate thoroughly with dilute HNO_3 (2 per cent). Now set the funnel in the flask and dissolve the precipitate back into it with dilute ammonia (1:4) using not to exceed 30cc. Finally wash the filter, using as little water as possible. To save time some chemists puncture the filter and wash the precipitate through with water. Then wash the funnel with ammonia, then with water. Now add 80cc of dilute H_2SO_4 (one part in four), and then ten grammes of pulverized zinc, *which must be free from much iron.* Now warm till rapid effervescence ensues and heat gently *ten minutes.* At the end of this time reduction will be complete. Meanwhile fold a 12 c. m. filter in "ribs;" put it in a funnel, and as soon as the reduction of the MoO_3 is complete pour the liquid off of the residue of the zinc into the filter, receiving the filtrate in a white china dish, rinse the flask and zinc once with water, pour this on the funnel after the liquid has run through, then fill up once with water and let that run through. Thin filter paper must be used, so that the whole operation of filtration and washing the zinc shall occupy but three or four minutes.

Now run the permanganate into the dark colored filtrate till the color is discharged, and the last drop gives a faint pink tint, marking the close of the reaction.

There is always some impurity in the zinc, hence it is absolutely essential, to make a *blank test* using the 30cc of NH_4HO, the 10 grms. of zinc and 80cc of sulphuric acid as before but omitting the yellow precipitate. The filtrate in this case will always require a small amount of permanganate, which must be determined, and deducted from the amount consumed in the regular determination, the difference being the permanganate solution equivalent to the yellow precipitate.

The number of cubic centimeters of permanganate solution used, after correction for error of standard, divided by the number of grammes of metal taken will give the amount of phosphorus in hundredths of a per cent.

In working this process it is important to check it from time to time upon material similar to that to be analyzed, and in which the phosphorus has been determined gravimetrically.

Notes on Above.—This is the original process of Emmerton and has been the starting point of a large number of modifications which increase the speed. The whole process depends upon such a nice adjustment of conditions that any of these methods before being adopted must be worked by a number of chemists and tried by much experience. Some of them are very highly recommended and are undoubtedly accurate where used, whether they would be safe to adopt generally more experience must determine. The above process has been in use some years and proved satisfactory, still it should be carefully checked against standard gravimetric methods, from time to time.

Among the important modifications advocated are the following :

1. Use of wet methods of oxidation to save baking.

Drown.—Trans. P. Inst. Min. Engrs., vol. XVIII, p. 90—uses permanganate and HNO_3 of 1.135 Sp. gr., which dissolves silica. Does not evaporate to dryness.

Shimer.—Trans. Inst. Min. Engrs., vol. XVII, p. 100—uses permanganate with sulphuric acid. Evaporating until the HNO_3 is expelled.

Clemens Jones.—Trans. Inst. Min. Engrs., vol. XVIII, p. 705—uses the method of Drown, but slightly modified, also washes the yellow precipitate with $(NH_4)_2SO_4$ solution to avoid the presence of nitrates.

Babbitt.—Jour. An. and App. Chem., vol. VII, p. 165—advocates the temperature of 25°C instead of 85°, to prevent the precipitation of arsenic.

Clemens Jones Jour. Trans. Inst. Min. Engs., vol. 17, p. 411, performs the reduction of the MoO_3 by filtration through zinc.

Other papers of importance are by Cheever, Trans. Inst. Min. Engs., vol. 14, p. 372, and a note from Stahl u. Eisen in Jour. Soc. Chem. Inds., vol. 11, p. 845.

The titration process may be applied to ores. These should be dissolved in HCl. Care must be taken to destroy all organic matter, as this may adhere to the yellow precipitate and cause reduction of the permanganate.

Evaporate the HCl solution with HNO_3, bake and then follow Emmerton. The writer's experience has been that, while good results were obtained with many ores, with some the process seemed to fail.

The titration method may be applied to estimating the yellow precipitate obtained as described in the direct weighing process, but the "shake down" method is the most rapid.

Titration of the yellow precipitate by standard alkali.—This method appears good for iron containing only small percentages of phosphorus. It may be found described in the following papers:

Jour. An. and App. Chem., vol. VI, p. 82.

Jour. An. and App. Chem., vol. VI, p. 204.

Jour. An. and App. Chem., vol. VI, p. 242.

Zeitschrift fur Anal. Chem., vol. XXVIII, p. 171.

A "mechanical" process has been used by which the yellow precipitate is separated in a centrifugal machine and its amount measured in a graduated tube. The results can hardly be very exact. See Jour. An. and App. Chem., vol. IV. p. 13.

THE DETERMINATION OF SILICON IN IRON.

The metals in which silicon has most frequently to be determined are pig iron, containing from one-half to four or five per cent. "ferro silicon," containing up to 14 per cent. or more, steel, with from traces to one-fourth per cent., and wrought iron with small fractions of a per cent.

In all these the silicon exists as Si, not as SiO_2, though there may be a little SiO_2 included as intermixed slag, especially in wrought iron.

All of these are easily soluble in HNO_3, 1.2 sp. gr., except ferro silicon, the Si being oxidized to SiO_2, part of which passes into solution. Evaporation of the HNO_3 solution to dryness, baking and re-solution in HCl only partially renders this SiO_2 insoluble.

To accomplish the complete separation of the SiO_2 by this means it is necessary to dissolve the metal in 1.2 sp. gr. HNO_3. Evaporate to dryness, bake as in the phosphorus determination, dissolve in HCl, and again *evaporate to complete dryness*, expelling all the HCl. On taking up again in HCl the SiO_2 is all left insoluble. After dilution the solution may be filtered from the residue of $SiO_2 + C$. This, after thorough washing first with HCl and then with water, may be ignited till the carbon is burned off, and weighed.

The SiO_2 thus obtained is never pure, and must be treated by H_2SO_4 and HFl, or must be fused and the SiO_2 determined. (See analysis of limestones.)

Hydrochloric acid or aqua regia may be used in place of HNO_3, but they do not attack ordinary iron so rapidly. Finally, solution in H_2SO_4 and evaporation till fumes of H_2SO_4 are given off has been used.

For details of these various methods see—

Blair, Chem. Anal. Iron and Steel 2d Ed., p. 72, nitric acid method.
Troilus, Notes on Chem. of Iron, p. 35, sulphuric acid method.
Also Trans. Inst. Mining Engineers, vol. X, p. 162 et seq and 187 et seq.

When a solution containing SiO_2 is evaporated with H_2SO_4 this will expel all volatile acids, and if the temperature is finally raised to near the boiling point of the concentrated acid, the silica is completely dehydrated and becomes insoluble. Titanic acid if present passes into solution and the silica thus obtained is pure. The following method, slightly modified from one published by Dr. Drown, depends upon this fact.

Trans. Am. Inst. Min. Engrs., vol. VII, p. 346.

In preparing the drillings for analysis great care must be taken to keep them free from sand. This is very difficult in the case of pig iron, hence it is always safer to clean them.

This is easily accomplished by folding a sheet of paper over a magnet, then picking up the metal against the paper. The sand and other foreign particles are left behind. On drawing the magnet away from the paper the drillings will fall off and can be collected on a clean sheet of paper. All the drillings must be gone over and no considerable residue should remain, if much graphite-like substance is separated it may hold silicon belonging in the sample.

The drillings should be fine, large fragments of metal dissolved slowly and may be left as hard grains in the silica, of course vitiating the result.

If these lumps remain, add more acid and heat slowly until they dissolve.

In "weighing out" great care must be taken to secure an average of fine and course, as these usually differ in per cent. of silicon.

Process for Pig Iron and Steel.—Weigh 1 grm. of pig iron or 5 grms. of wrought iron or steel. Put into a caserole or dish and cover with a large watch glass. Add carefully 30cc of a cold mixture of 8 parts by volume of conc. HNO_3, 5 parts of conc. H_2SO_4 and 17 parts of H_2O (for 1 grm. of pig iron) or 100cc of a mixture of 35 parts of conc. HNO_3, 15 parts of H_2SO_4 and 50 parts of H_2O (for 5 grms. of steel).

Warm till action ceases, then boil down rapidly on an iron plate or over the bare flame until the $Fe_2(SO_4)_3$ separates as a white mass; continue the heating until dense fumes of H_2SO_4 are evolved. These have a peculiar suffocating odor, easily recognized; their formation indicates the total expulsion of the HNO_3. This is absolutely necessary in order to make the silica insoluble. There will be danger of "spattering" unless the heating be carefully done, but if the dish be well covered this need cause no loss.

Now let cool, and then wash off the cover into the dish. Dilute to 150 or 200cc, cover, set over a lamp and boil until all $Fe_2(SO_4)_3$ is dissolved, as can be recognized by the disappearance of the silky precipitate in the liquid. Continue the boiling for five minutes. Wash off the cover, then let the liquid stand until all the SiO_2 settles. Decant the clear liquid through a 7 c. m. ashless filter, previously washed out with boiling water. Finally transfer and wash the residue with hot water. When partially washed, drop a little HCl on the filter and residue then wash again with hot water till the filtrate no longer tastes acid. Without drying transfer the filter to a crucible and ignite, gently at first, finally at high heat, until all the carbon (graphite) is burned and the SiO_2 is white. If this be done in a platinum crucible and over a blast lamp the "burning off" of the carbon need not take more than a few moments. It may be much hastened by directing a very gentle current of oxygen gas into the crucible. If care be taken not to blow out any of the light particles of SiO_2, this is a good plan. The weight of the SiO_2, multiplied by 0.4667, gives the Si.

If the above directions are exactly followed as to diluting and boiling the solution after evaporation, there will be no need of a filter pump to secure rapid filtration. Boiling with a large excess of water consolidates the SiO_2 so it filters easily.

In the case of steels, the SiO_2 being very small in amount, it is necessary to test its purity. Add a drop of H_2SO_4, and then either a few drops of pure HFl or a few crystals of pure NH_4 Fl. Evaporate to dryness and ignite strongly. The SiO_2 goes off as gaseous $SiFl_4$. The residue, if any remains, is to be deducted from the total weight, the difference being SiO_2.

At the Edgar Thomson Steel Works a special process for Si in pig iron is used. They chill the iron in water which makes it brittle. This is then pulverized in a steel mortar, dissolved in HCl, rapidly evaporated to dryness, taken up in HCl, diluted and filtered. Without drying, the residue is put into a platinum crucible, ignited in a stream of oxygen and weighed. The time is said to be 12 minutes. Of course it is not very exact.

For details see Blair Chem. Anal. iron and steel; second edition, p. 77.

Determination of Silicon in Ferro Silicon.—This material is not easily attacked by any of the above mixtures. It can usually be dissolved by prolonged boiling with aqua regia, adding fresh acid from time to time. Finally add 25cc of dilute (1:3) H_2SO_4, evaporate until fumes of SO_3 appear, and then finish as in the regular process.

Those samples which aqua regia will not dissolve are, according to Williams, best treated by fusing with 6 or 8 times the weight of dry Na_2CO_3. Then proceeding with the fusion, as in the determination of SiO_2 in the insoluble matter of a limestone. The metal must be very finely pulverized and not more than 0.5 grm. taken.

Williams. Trans. Am. Inst. Min. Engrs., vol. XVII, p. 542.

THE DETERMINATION OF MANGANESE.

Two classes of substance present themselves. First, ores, slags and metals high in manganese, such as manganese ore and ferro manganese, with from 15 to 90 per cent. of manganese; second, ordinary iron ores, irons and steels, with from a trace to about 3 per cent. of manganese.

THE ACETATE PROCESS, FOR THE DETERMINATION OF MANGANESE IN ORES WITH HIGH PERCENTAGES.

The process depends upon the separation of the iron and alumina as basic acetates, precipitation of the Mn by bromine, re-solution and determination as pyrophosphate.

When a solution of Fe_2Cl_6 and $MnCl_2$ is boiled with sodium or ammonium acetate, the iron is precipitated completely, provided, first, the solution is sufficiently dilute, containing less than one gram. Fe_2O_3

in 500cc; second, the amount of free acid is very small. This precipitate is nearly free from manganese, provided the excess of acetate of soda is very small. If bromine water is added to the filtrate, the Mn is completely precipitated as MnO_2, provided a considerable excess of sodium acetate is present. If ammonium salts are present, the MnO_2 will only separate when the solution is made alkaline. The iron solution must contain no *ferrous salt* or a red "brick dust" like, slimy precipitate will form, and the filtrate will be cloudy and deposit iron.

The process is perfectly satisfactory provided all details are very carefully attended to.

Process for Ores.—Dissolve one-half grm. of the ore in 15cc concentrated HCl, dilute and filter as in the iron assay.

Evaporation to dryness is usually unnecessary, few ores containing soluble silicates. When such occur, as in slags, dissolve in dilute acid, evaporate to dryness, add HCl and then water.

If no chlorine is given off when the ore is dissolved, owing to the absence of MnO_2, ferrous iron may be present. In this case add a crystal of $KClO_3$ and boil until all the Cl is expelled.

When a ferric chloride solution is evaporated to dryness in the presence of organic matter a slight reduction to ferrous salt often occurs; hence, in this case always oxidize the solution after filtration by adding a little $KClO_3$ or HNO_3. The solution must be boiled till all the Cl is expelled or Mn will precipitate with the iron in the subsequent separation.

See Blair Chem. An. Iron, page 227.

To the filtrate add a solution of sodium carbonate carefully until a slight permanent precipitate forms. Redissolve this with a few drops of HCl, giving each drop two or three minutes to act, and stopping as soon as the solution clears.

Now dilute to about 300cc, add one grm. sodium acetate, cover the beaker and boil vigorously till the iron separates. Should it not come down promptly drop in a solution of Na_2CO_3 drop by drop until the precipitation is complete. The liquid tested by a slip of litmus paper must be distinctly acid. Let the precipitate settle clear and decant the liquid through a 9 c. m. filter as close as possible, add 150cc of boiling water, settle, pour off and finally run the precipitate on to the filter and wash once with hot water. Wash it off the filter back into the beaker. Dissolve it in its least possible quantity of HCl and repeat the precipitation exactly as before. Transfer the precipitate to the filter and wash well with *hot water*.

Test this last precipitate for Mn by fusing with Na_2CO_3 and $NaNO_3$ on a platinum wire. If not free from Mn a third precipitation will be necessary.

The filtrate will amount to about a litre. It should be perfectly clear and colorless. Concentrate it to about 500cc, then add ten grms. of sodium acetate and an excess of bromine water. Warm until the MnO_2 has settled and the liquid is clear. Filter on to a 7 c. m. filter and wash well with hot water.

Wash the precipitate off the filter into a beaker. Now wash the filter paper with dilute HCl, in which a small crystal of oxalic acid is dissolved. This will dissolve off all adhering MnO_2. Heat the HCl containing the MnO_2 in the beaker, adding oxalic acid solution drop by drop until the MnO_2 is all dissolved. Now dilute to 150cc, and add NH_4HO till a slight permanent precipitate is formed, then a few drops of acetic acid and boil. If any precipitate of Fe_2O_3 separates filter it off and wash it.

Unless this is light red in color and very small in amount, redissolve it in a few drops of HCl, add H_2O, then NH_4HO, then acetic acid as before. Boil till the precipitate separates and filter into the original solution. This re-solution is essential in most cases, and need delay the work but a few moments. The object of the oxalic acid is to reduce the MnO_2 to MnO, and so make it dissolve quickly. MnO_2 is very slowly attacked by dilute HCl.

To the filtrate, now perfectly clear and colorless, add an excess of a solution of microcosmic salt ($Na_2NH_4PO_4$). Now heat to boiling, add NH_4HO drop by drop as fast as the precipitate formed by each addition becomes " silky " in appearance, stirring all the time to prevent bumping. When no more precipitate forms add enough NH_4HO to make the solution smell slightly of NH_3, and boil till the precipitate is completely silky and settles quickly. Now cool the liquid and filter. Wash the precipitate with water containing a few drops of NH_4HO. Ignite, weigh as $Mn_2P_2O_7$, containing 0.3874 manganese.

The acetate of soda used must be tested for Mn. If any is found, dissolve the salt in water, add bromine water and boil till all the bromine is expelled. Filter the solution from the MnO_2 thus separated and use it instead of the solid salt.

Process for Spiegle Iron and Ferro Manganese.—Take one-half gram. of the drillings, dissolve in 10cc HNO_3 1.2 sp. gr., evaporate to dryness and "bake," then dissolve in 10cc conc. HCl, add a little bromine water (to reoxidize any FeO formed), boil down till all excess of bromine is gone and most of the HCl evaporated. Dilute, filter if necessary; precipitate the iron and determine the Mn as by the method for ores.

Some ores and spiegle irons contain copper and nickel. These will come down with the MnO_2 in part at least. They should be separated by H_2S, which will precipitate Ni. Co. Cu. and Zn., but not Mn from a solution containing a slight excess of acetic acid.

This may be done in the original acetate filtrate. The solution must be boiled till all H_2S is expelled before adding bromine.

THE ACETATE PROCESS FOR IRON ORES CONTAINING BUT LITTLE MANGANESE.

In this case it is desirable to work upon larger amounts of material. The filtration and washing of a large basic acetate precipitate is very troublesome, and can be avoided by taking an aliquot part of the solution after the precipitate has settled.

The error introduced by the bulk of the precipitate is inappreciable when the per cent. of manganese is small.

A single precipitation of the iron is entirely sufficient, provided care be taken to avoid excess of sodium acetate.

Extreme care in measuring the solution, as well as in keeping the temperature constant, is also superfluous where under three per cent. of manganese is present and the volumes are kept large, as 10cc on a litre could only cause an error of 0.03 per cent.

The precipitate by bromine is MnO_2. On ignition it changes to Mn_3O_4, but only when ignited under very exact conditions as to temperature and access of air. This precipitate also usually retains small amounts of soda salts. For these reasons the percentage of Mn it contains will always be a little uncertain.

However, as these variations are limited to a small percentage of the weight of the precipitate, the results obtained by direct weighing, where but little Mn is present, will be sufficiently accurate for all ordinary work.

Process.—Dissolve four grms. of the ore in 30cc conc. HCl exactly as in the iron assay. If there is any ferrous iron present add about 1cc of HNO_3 to completely convert it to ferric chloride. Boil the solution until all chlorine and excess of HNO_3 are expelled. The evaporation must not go so far that insoluble iron salts separate; should such form add more HCl and heat until they dissolve. Add water, warm and filter from the residue.

Take a large Erlenmeyer flask, one which will hold when quite full 2400cc. Dry it, then measure into it exactly 2000cc of water; this should reach up into the narrower portion of the flask. Paste a thin strip of paper on the glass to exactly indicate the level of the liquid. The flask must be set on a level desk, and the place it stands as well as the position of the paper mark noted, so that it can be subsequently returned *to the same position.*

Now transfer the solution of the ore to the flask and dilute it to about 1700cc. Then add a solution of Na_2CO_3 gradually until the liquid begins to grow dark red. Continue to add the reagent drop by drop, shaking the flask after each addition until the liquid is very dark in color and the precipitate formed only redissolves slowly. The object is to reach a point just short of that at which the iron will be precipitated. The operation requires practice. Should the point be overstepped, add a little HCl, and when the liquid becomes clear, neutralize over again; but in this case, in the writer's opinion, the iron precipitate is more liable to contain Mn.

Now add six grms. of pure sodium acetate. Set the flask on a hot plate and boil the solution hard.

The iron should immediately separate as a bulky, red precipitate. If it fails to do so at once drop in very cautiously a dilute solution of Na_2CO_3 until the separation is complete. Now boil a few minutes longer, then remove the flask to the place where it stood when graduated, placing it in the same position, and fill it exactly to the mark with cold water. With a long rod stir the liquid thoroughly, then let it settle. As soon as it is clear pour off one litre into a graduated flask. This whole operation can be done so quickly that the liquid will not cool materially.

Filter the measured portion of the liquid. The filtrate should be colorless and distinctly acid to litmus paper.

Concentrate the filtrate to about 500cc. Add five grms. of sodium acetate and boil. Should any precipitate form filter it off, dissolve it in HCl containing a little oxalic acid, add a solution of Na_2CO_3 until a slight permanent precipitate forms, then acetic acid till just acid. Boil this liquid, filter from any precipitate, and add the filtrate to the main solution.

Finally add bromine water, warm until the MnO_2 settles completely, filter, wash well with hot water, ignite and weigh as Mn_3O_4 containing 0.7205 Mn. Calculate the result as though two grms. of ore had been taken.

Should the ore leave little residue it need not be filtered off, but may go into the flask with the solution. In applying this process to slags and ores containing decomposable silicates, the HCl solution must be evaporated to dryness, taken up again in HCl, HNO_3 added and boiled off as usual.

If care be taken in the neutralizing no precipitate will form on concentrating the filtrate from the iron and delay will be avoided.

Should the ore contain nickel or copper, these will contaminate the manganese precipitate and the results will be inaccurate. In this case the precipitate of MnO_2 must be redissolved in HCl containing a little sodium sulphite. The solution boiled till free from SO_2 and then cooled and nearly neutralized by Na_2CO_3, a little sodium acetate added and the Ni and Cu precipitated by H_2S. In the filtrate from these sulphides the Mn can be determined either by precipitation with bromine or as phosphate.

The Acetate Process may be Applied to Pig Iron and Steel. — Dissolve 4 grms. in 50cc HNO_3 1.2 sp. gr., add 10cc conc. HCl. Evaporate to dryness and "bake." Redissolve in 25cc conc. HCl, add a little HNO_3 boil and proceed as with ores low in manganese. Filtration from the insoluble residue is unnecessary in this case.

References on the acetate process :

Blair—Chem. Anal. Iron, Second Ed., p. 103, et. seq., also p. 227.
Jour. Soc. Chem. Indst., vol X, p 101, on the properties of Mn_3O_4.
Trans. Inst. Min. Engrs., vol X, p. 101.
Gibbs—Sillimans Am. Jour. [11] 44, p. 216, on the determination as phosphate.

THE POTASSIUM CHLORATE, OR FORD—WILLIAMS' METHOD FOR MANGANESE.

This is a volumetric method depending upon the precipitation of the Mn as MnO_2 by $KClO_3$ from a solution in concentrated HNO_3, and after filtering off and washing the MnO_2 determining it volumetrically by measuring its oxidizing power.

See Trans. Am. Inst. Min. Engrs., vol. IX, p. 397—Ford.
Trans. Am. Inst. Min. Engrs., vol. X, p. 100—Williams.
Trans. Am. Inst. Min. Engrs., vol. XII, p. 73—Troilus.
Trans. Am. Inst. Min. Engrs., vol. XIV, p. 372—Cheever.

The method is especially adapted to the determination of manganese in steels and irons low in silicon and dissolving in HNO_3 without residue.

When $KClO_3$ is added in successive small portions to a boiling hot solution of Mn in concentrated HNO_3, the Mn is completely and rapidly precipitated as MnO_2, provided an amount of iron at least equal to the Mn, be present, HCl be absent and the HNO_3 in sufficiently large excess and sufficiently concentrated.

This MnO_2 is entirely insoluble in cold concentrated HNO_3 provided this contains no lower oxides of nitrogen ("red fumes"); if these are present, that is if the HNO_3 is not *perfectly colorless*, the MnO_2 will be reduced and dissolved.

The precipitate contains a little iron but is free from other impurities.

When the solution to be precipitated by $KClO_3$ contains any HCl, this will be first acted upon and destroyed before the MnO_2 will separate. Chlorine being driven off and water formed by the oxidation. This will result in a weakening of the HNO_3, and hence in this case more HNO_3 must be present to prevent too great loss of strength.

SiO_2 in the solution may separate in a gelatinous form and prevent filtration, hence it must always be first removed.

Chlorate Process for Steel Low in Silicon — Precipitation of the MnO_2.— Dissolve 5 grms. in 60cc 1.2 HNO_3, in a No. 2 beaker. Evaporate down to 25cc, then add 100cc of *colorless conc.* HNO_3. Set on an iron plate and heat to incipient boiling. Now drop in powdered $KClO_3$, a little at a time, adding each portion when the effervescence from the preceding portion has ceased. By the time 2 to 2½ grms. have been added the MnO_2 will have separated as a fine brown powder. Now add ½ grm. more $KClO_3$ and boil gently for 10 minutes. Then add 1 grm. more $KClO_3$ and 25cc conc. HNO_3 and boil 10 minutes longer. Remove from the plate and cool by setting in water. When the MnO_2

has settled, filter without dilution, through an asbestos filter,[1] finally run the MnO_2 on the filter and wash beaker and filter with *colorless concentrated* HNO_3 three or four times. This can be done without using more than 15 or 20cc, adding only a little each time and letting each portion run through before adding the next. Finally wash with a little cold water. If the HNO_3 is colored by lower oxides of nitrogen (from standing and the action of light), it may be purified by blowing a strong current of air through it for some time, until colorless.

After washing the MnO_2 with cold water till the acid taste is gone from the filtrate, (letting each successive portion of water run entirely through before adding the next, so as not to use in all more than 20cc) wash the asbestos and precipitate back into the beaker (which will always have some MnO_2 adhering to it), and proceed with the volumetric determination of the manganese.

Volumetric Determination of the MnO_2.—
This process consists in dissolving the MnO_2 in a measured excess of an acid solution of ferrous sulphate of a known strength. Each molecule of MnO_2 changes two molecules of ferrous sulphate to ferric sulphate. The amount of ferrous sulphate remaining is then determined by a standard solution of potassium permanganate. The reactions are as follows:

1. $MnO_2 + 2 FeSO_4 + 2 H_2SO_4 = MnSO_4 + Fe_2(SO_4)_3 + 2 H_2O.$
2. $10 FeSO_4 + K_2 Mn_2 O_8 + 8 H_2SO_4 = 5 Fe_2(SO_4)_3 + K_2SO_4 + 2 MnSO_4 + 8 H_2O.$

In all work involving the use of potassium permanganate only glass stoppered or Gay Lussac burettes must be used, as it is reduced and destroyed by contact with all organic materials, such as rubber tubes, paper, etc. There are required, first, a solution of potassium permanganate of a known strength; second, a solution of ferrous sulphate in dilute sulphuric acid. The strength of this is determined by titration with the permanganate solution.

Preparation of the Permanganate Solution.—Dissolve 1.149 grms. of pure potassium permanganate in water and

---

1. *To Prepare the Asbestos Filter.*—Melt the bottom of a six-inch test tube and draw it out into a narrow tube. Cut off the end and into the long funnel thus formed drop a little disc of platinum foil punched full of holes and fastened to a wire which can run into the funnel stem and hold the foil in place. Then put in a little asbestos which has been boiled with HCl, washed and finally ignited in a platinum crucible. Do not "pack" it in, simply pour on water and let it run through so as to settle it on the bottom. By a little practice a filter can be made in this way which will work rapidly and yet retain all the MnO_2.

dilute to one litre. 1cc of this solution will have the same oxidizing power as 0.001 gram. of manganese in the form of the brown precipitate (MnO_2). Check the solution against pure iron or pure ammonium ferrous sulphate $(NH_4)_2$ Fe $(SO_4)_2$ 6 H_2O. Dissolve 0.1425 grms. of the salt in 50cc of water containing 2cc of H_2SO_4. This should consume just 10cc of the permanganate solution. Run in the solution until the last drop gives a permanent pink color.

If more or less than 10cc is required, calculate the amount of Mn to which each cc of the permanganate is equivalent by the proportion.

.001 : x = n : 10, n being the number of cc of solution used in the test, and x the required value.

Preparation of the Ferrous Sulphate Solution.—Dissolve 20.18 grms. of pure crystallized ferrous sulphate ($FeSO_4 7H_2O$) in about 500cc of water, to which 25cc of concentrated H_2SO_4 has been added, and then dilute to one litre.

Determine its strength against the permanganate solution by measuring 5cc with a pipette into a beaker, adding about 25cc of water and then running in the permanganate till the pink color is permanent. About 10cc should be required.

This value must be redetermined frequently as the solution of ferrous sulphate alters rapidly from the oxidizing action of the air.

In a large way it is best kept in a carboy and covered with a layer of kerosene oil to keep out air.

The solution can be drawn out by a siphon, and when used in this way alters less rapidly.

From the two formulas already given we have the relations between the MnO_2, $FeSO_4$ and $K_2Mn_2O_8$ as follows:

One atom of Mn in the form of brown precipitate (MnO_2) will oxidize 2 atoms of Fe as ferrous sulphate. 1 molecule of permanganate will oxidize 10 atoms of Fe as ferrous sulphate, that is to say one molecule of permanganate will oxidize the same amount of iron as will 5 molecules of MnO_2 containing 5 atoms of manganese.

Therefore to find how much $K_2Mn_2O_8$ will be needed to have the same oxidizing power as 0.001 grms. of Mn in the form of the brown precipitate we have the proportion.

Wt. 5 atoms Mn : wt. 1 mol. $K_2Mn_2O_8 = 275 : 316 = .001 : x$, which gives $x = .001149$ grms., the amount of $K_2Mn_2O_8$ to be dissolved in 1cc if 1cc is to be equivalent to .001 grm. Mn as "brown precipitate." This is 1.149 grms. in a litre.

To determine the amount of iron, or of ammonium ferrous sulphate to which 1cc should be equivalent, we have,

Wt. 1 atom Mn : Wt. 2 atoms Fe $= 55 : 112 = .001 : x$, in which x is the required amount of iron. The value of x is 0.002034. To determine the amount of the am. fer. sulph., as this contains $\frac{1}{7}$ of its weight of iron, multiply the value of x by seven $\therefore = .01425$ for 1cc, or the figure given in the directions for making the check for 10cc.

That 5cc of the ferrous sulphate solution may be equivalent to 10cc of the permanganate it must contain 0.02034 Fe. This corresponds to 20.18 grms. of $FeSO_3 \, 7H_2O$ to the litre.

Determination of the MnO_2.—To the asbestos and MnO_2 in the beaker, add the solution of ferrous sulphate from a pipette 5cc at a time until, after stirring and warming, the MnO_2 is *completely* dissolved. It is best to take the same pipette used in standardizing. Break up all lumps of asbestos or precipitate with a glass rod as they may conceal undissolved particles of MnO_2. Now add a little water and run in the permanganate solution till a permanent pink color is produced (it may disappear in two or three minutes, but this is of no consequence). Read the burette and deduct the amount used from that to which the amount of ferrous sulphate taken would have been equivalent — the difference is that equivalent to the Mn present in the precipitate. This, corrected by the factor for the permanganate solution will give the amount of Mn in milligrams.

As an example—Suppose that 5cc of ferrous sulphate solution equaled 9.6cc of permanganate solution, and 10.3cc permanganate equaled 0.1425 grms. of ammonium ferrous sulphate. If 15cc of ferrous sulphate solution was added to dissolve the MnO_2 and the permanganate required to oxidize the excess was 4.5cc, then the calculation would be as follows:

$3 \times 9.6 = 28.8 =$ the permanganate equivalent to the fer. sulph. used.
$4.5 =$ the "Titre back."
$\overline{24.3} =$ the number of cc of permanganate equivalent to the precipitate.

$24.3 : x = 10.3 : 10$ (x being the true amount of correct permanganate).
$x = 23.6 = 0.0236$ grms. Mn in the precipitate.

THE CHLORATE PROCESS FOR ORES.

Take 5 grms. dissolve in 50cc of conc. HCl. Evaporate to dryness, avoiding a temperature above 100°, add 20cc HCl, and then water. When dissolved filter into a No. 2 beaker. Add 50cc conc. HNO_3, evaporate to a syrup, then add 100cc of conc. HNO_3 and proceed as before.

THE CHLORATE PROCESS FOR PIG IRON.

Dissolve 5 grms. of the metal in HNO_3 1.2 sp. gr., taking about 60cc. Then add 25cc HCl, evaporate to dryness and bake. Dissolve in HCl, filter from the SiO_2, and to the filtrate add 0.2 grm. ammonium fluoride or a few drops of hydrofluoric acid. Then add 50cc of HNO_3. Concentrate to a syrup, add 100cc HNO_3 and proceed as before. The hydrofluoric acid expels traces of SiO_2 from the solution and greatly accelerates the filtration from the MnO_2. (E. F. Wood.)

The above process, carefully conducted, gives very accurate results, but it should be checked against metals in which the Mn has been carefully determined by the acetate method until the two work together.

For Ferro Silicon and Ores high in Manganese.—The permanganate solution should be standardized by working on a metal of known percentage of Mn, as the composition of the precipitate is considered by some chemists not to be exactly MnO_2, and by thus standardizing in the same way that the ore is analyzed all risk from this source is avoided. Where only small amounts of Mn are present this source of error is unimportant.

The acetate process is probably better adapted to all high manganese materials, and if skillfully worked is nearly as rapid as the other when HCl has to be expelled by evaporation with HNO_3.

One slight objection to the chlorate process is the large amounts of expensive acids required.

To avoid this the process can be worked on smaller amounts of substance, but great care and skill is then needed to secure close results. All measurements and titrations must be very exact.

Take one grm. of iron or steel, dissolve it in 15cc of HNO_3 1.2 sp. gr. Evaporate to 10cc, add 35cc HNO_3, and after precipitation and boiling 10cc more, cutting down the $KClO_3$ to about one grm., filter, wash and proceed with the volumetric determination of the precipitate.

COLOR PROCESS FOR MANGANESE.

This depends upon the change of Mn to permanganic acid when boiled in nitric acid solution with PbO_2.

It is sufficiently accurate for steels, irons and ores when the amount of Mn is small, provided exactly the same manipulation is adhered to in every case.

It requires, first, a standard material containing a known amount of Mn, of approximately the same per cent. as the material to be tested, and of *precisely the same kind* (steel for steel and pig iron for pig iron, ore for ore, etc.). Second, PbO_2 *free from Mn*.

The method is usually restricted to steels, but may be used with certain modifications for an approximate estimate of the Mn in ores and pig iron containing very small amounts.

Process for Steel.—Dissolve 0.2 grms. of steel in a test tube in 15cc HNO_3 1.2 sp. gr. Cover with a small glass bulb and heat in a water bath until solution is complete. Now filter if necessary, dilute to 100cc and mix. Put 10cc of this solution in an 8-inch test tube, add 3cc HNO_3 1.2 sp. gr., heat till the solution boils rapidly. (This is best done by setting the tube in a $CaCl_2$ bath, boiling at 115°C.) Now add carefully while boiling 0.5 grms. PbO_2 and continue to boil for *exactly five minutes*. The tube must be covered all the time by a little glass bulb, or dry iron salts will form on the sides.

Now set the tube vertically in cold water and let settle till the PbO_2 is all on the bottom and the violet liquid is absolutely clear. Avoid exposure to bright light, which may change the color. There must have been prepared at the same time and in the same way a solution of the standard steel. Pour off the two solutions into two "carbon tubes" and dilute till the color "matches." The volumes will then have the same ratio as the amounts of manganese in the standard and the test sample.

To Apply this Process to Pig Iron.—Dissolve 0.2 grms. in 5cc of HCl of 1.1 sp. gr. (1:2) and filter the solution, evaporate it to a syrup with 10cc conc. HNO_3, add 5cc more, dilute and proceed as with steel.

In the Case of Ores.—Dissolve 0.2 grm. in 5cc of HCl and boil to a syrup, add 10cc conc. HNO_3, evaporate to a syrup, add 10cc HNO_3 1.2 sp. gr., dilute, filter and proceed as before.

References on the Color Process—

Blair Chem. Anal. Iron and Steel, 2d Ed., p. 120.
Hunt Trans. Am. Dist. Min. Engrs., vol. 15, p. 164.

Numerous other methods are in use for Mn. Chief among these is Volhard's method, titrating the Mn in the filtrate after separating the iron by ZnO.

See Blair Chem. Anal., 2d Ed., p. 112.

The titration is difficult except for an expert.

THE DETERMINATION OF SULPHUR.

Sulphur occurs in iron ores, both in the form of sulphides, such as pyrites (FeS_2) and as sulphates, such as gypsum ($CaSO_4 2H_2O$), or occasionally barite ($BaSO_4$).

In iron and steel the sulphur is entirely in the form of sulphide.

In all gravimetric methods the sulphur is first entirely converted into some soluble sulphate, and then determined by precipitation with barium chloride as barium sulphate ($BaSO_4$).

In order to be satisfactory the precipitation must be done under carefully regulated conditions.

Barium sulphate, while entirely insoluble in water, is not so in dilute acids, the amount dissolved increasing with the concentration of the acid. The presence of a considerable excess of $BaCl_2$ in the liquid appears, however, to *counteract* this solvent action. Thus a sufficient excess of $BaCl_2$ will completely precipitate the sulphuric acid from a solution strongly acid with HCl.

When $BaSO_4$ is precipitated in solutions containing much iron, basic iron salts will adhere to the precipitate, making it reddish in color unless the solution contains considerable excess of HCl. This color is shown best after the precipitate is ignited. Some of the sulphuric acid appears to be in combination with this iron, and is driven off on ignition, leaving ferric oxide; so that these impure precipitates are liable to give *low results*, particularly if subsequently purified and the $BaSO_4$ they contain determined.

Barium salts, particularly the nitrate ($Ba(NO_3)_2$) have a strong tendency to adhere to the precipitate, making it impure; these cannot be completely washed out with water.

Acid solutions of Fe_2Cl_6 when hot, hold a little $BaSO_4$ in solution, this separates completely when the liquid cools. The precipitate of $BaSO_4$ is fine, liable to run through the filter and impure, when precipitated cold, or in concentrated solutions or by *too concentrated a solution of $BaCl_2$*. This last point is especially important. The solution of $BaCl_2$ must always be *well diluted* and heated to boiling before being added. Working in this way will give a granular, rapidly subsiding precipitate easy to wash and *pure*.

See Fres. Quant. Anal., § 71 (a). Archbutt—Jour. Soc. Chem. Inds., 9, p. 25.
Lunge—Jour. Soc. Chem. Inds., 8, p. 967; also a note in the same Journal, 8, p. 819.

Conversion of the Sulphides to Sulphates.—All sulphides are completely oxidized to sulphates when fused with a mixture of dry Na_2CO_3 and $NaNO_3$.

As certain bisulphides give off sulphur vapor at a comparatively low temperature (below the fusing point of Na_2CO_3), when these or free sulphur are present care must be taken to prevent loss in this way. The mixture of ore and flux must be covered with a layer of pure "fusion mixture" and heated carefully.

After fusion all the sulphur, whether originally present as sulphides or sulphates (including $BaSO_4$) will be found as Na_2SO_4, the other bases remaining as oxides or carbonates. When the fused mass is boiled with water till thoroughly disintegrated and then filtered and washed the sulphur all passes into the filtrate.

Sulphides can be more or less completely oxidized to sulphates in the "wet way" by treating them with hot concentrated HNO_3 or aqua regia. These methods are not very satisfactory, as free sulphur is liable to separate and fuse in drops. Once in this form it resists boiling with all ordinary oxidizing agents for a long time. Iron sulphides can be completely oxidized by boiling with a large excess of *concentrated* HNO_3 and adding a little powdered $KClO_3$.

When iron sulphides, or even iron containing but little sulphur, is dissolved in dilute (1.2 sp. gr.) HNO_3 a considerable amount of the sulphur separates as such and escapes oxidation.

Solutions containing sulphuric acid, on evaporation to dryness and "baking," as is common in iron analysis, may loose SO_3 unless enough potash or soda be present to hold it all in combination, as the sulphates of iron are easily decomposed by heat.

See Fresenius — Quant. Anal. § 148, 2.
Blair — Chem Anal. Iron, p. 66 and p. 220.

Method for Sulphur in Iron Ores.—Mix one grm. of the finely pulverized ore with eight grms. of dry Na_2CO_3 and one-half to one grm. of $NaNO_3$, according to the amount of sulphur in the ore. Put the mixture in a platinum crucible and fuse carefully. As soon as it is well melted chill the crucible by dipping the bottom in water. Boil out the fusion with water until all the material is soft and no hard lumps remain, and if the solution is colored by $Mn_2Na_2O_8$, add a few drops of alcohol. Filter and wash well with hot water. Add HCl to the *filtrate* till just acid, and evaporate it to dryness carefully, and dry at 100°C. Now add 5cc of HCl first diluted with its own volume of water. Warm and add 50cc of H_2O, heat till all is dissolved but a little SiO_2, filter and wash. The filtrate should not exceed 100cc, if it does concentrate it. Now heat to boiling and add 5 to 10cc of a 10 per cent. solution of $BaCl_2$ previously diluted with 10 or 20cc of water and *heated*. Stir and let the precipitate of $BaSO_4$ settle. When clear, filter, wash with hot water, dry, ignite and weigh as $BaSO_4$, the weight of which multiplied by $0.137 = S$.

$BaSO_4$ is easily reduced to BaS by heating with carbon; hence, in igniting the precipitate detach it as far as possible from the filter, burn the paper carefully on a platinum wire, avoiding a high heat. Add the ash to the precipitate in the crucible and heat gently with the cover off until all carbon is burned, finally igniting to a bright red heat. After weighing the precipitate add a little water to it and test with turmeric paper. If it reacts alkaline the results are untrustworthy. Ordinary gas contains sulphur, which forms SO_2 on burning, and is liable to get into sulphur determination. Therefore, keep the crucible well covered during the fusion. In accurate work it is always necessary to make a blank analysis, and determine the small amount of sulphur contained by the reagents or absorbed from the gas flames. Deduct this from the amount found when working on the ore.

Method for Sulphur in Iron and Steel.—This is usually present in very small percentages.

Take from two to five grms., according to the per cent. of sulphur. Add from 25 to 40cc *concentrated* HNO_3. Cool the dish if the action is too strong, or heat it if too slow. When nearly all is dissolved, heat to boiling and add 2 or 3cc conc. HCl to complete the solution. Now add about one-half grm. $KClO_3$ free from sulphur. Boil to dryness and bake slightly. Add 10 to 20cc conc. HCl to dissolve it and again dry down thoroughly. Dissolve the residue in 15 to 40cc conc. HCl. Evaporate the solution until a skin begins to form on the solution or it becomes syrupy. Now add 5cc. of conc. HCl. When all dissolves clear dilute with its own volume of hot water and filter into a small beaker through a paper previously washed out with a little hot dilute HCl (this facilitates filtration), wash the dish and residue with hot water.

The filtrate and washings must not exceed 75cc. Now heat and add 5cc of a 10 per cent. solution of $BaCl_2$ and let stand till the liquid settles perfectly clear. (Two hours is sufficient if everything is right.)

Filter on to a small ashless filter, wash with water containing a little HCl, dry, ignite and weigh as $BaSO_4$.

The residue from which the solution for the determination of sulphur is filtered must contain no basic iron salts (due to the great concentration before dilution), as these may hold sulphur.

Jour. Anal. and App. Chem., vol. 6, p. 318.

This nitric acid method may be applied to ores. Use one grm. of ore, 20cc HNO_3 and a little $KClO_3$. Evaporate to dryness, take up in HCl and proceed as in the treatment of iron. This process will not determine the sulphur in any barium sulphate, or all that in any lead sulphide present, and hence is not so universally applicable as the regular fusion method.

THE DETERMINATION OF SULPHUR IN PIG IRON AND STEEL BY EVOLUTION AS H_2S.

The oxidation methods, while accurate, and the only ones that can be relied upon to certainly give the total sulphur in any material, are slow.

When iron containing sulphur is dissolved in HCl, the sulphur is evolved as H_2S. This gas can be absorbed by various substances and the sulphur determined volumetrically or gravimetrically.

Several very rapid methods are based on this reaction. They are especially adapted to pig iron and steel containing but small fractions of a per cent. of sulphur, and when properly conducted are accurate, but are subject to a number of causes of error which must be carefully guarded against.

H_2S is very easily decomposed by comparatively feeble oxidizing agents, water being formed and free sulphur deposited. *Prolonged* contact with air and sunlight, solutions of Fe_2Cl_6, traces of chlorine, all act on it in this way. Therefore all these must be avoided in the process. There is no necessity, however, of working in an atmosphere of hydrogen or CO_2 if the process be *rapidly conducted*. On the other hand slow evolution, or HCl containing traces of Cl or Fe_2Cl_6 will cause retention of sulphur in the residue from the solution.

Dilute HCl, according to the writer's experience, frequently fails to cause complete evolution of the sulphur as H_2S where the concentrated acid succeeds.

Most irons when dissolved rapidly in hot *concentrated* HCl free from oxidizing impurities leave a residue practically free from sulphur.

This residue should always be examined for sulphur, however, unless previous tests on the same kind of metal have shown it to be free, in which case this is unnecessary. The presence of copper or arsenic in the metal would of course have a tendency to hold sulphur behind, as their sulphides are not decomposed by HCl.

PERMANGANATE, OR DROWN'S METHOD.

This is the quickest and least troublesome of the gravimetric evolution methods. It is accurate for small percentages of sulphur. A solution of $K_2Mn_2O_8$ will completely oxidize a little H_2S to H_2SO_4, provided the volume of liquid is large, otherwise free sulphur will separate.

Process.—Take 5 or 10 grms. of the borings, put them in a flask of 500cc capacity provided with a funnel tube having a stopcock, and a delivery tube. This latter should have a bulb blown in it and be so inclined that any acid condensing in the tube may run back into the flask. It is a good plan to jacket this tube with a larger one, filling the space between them with water to cool it and prevent free acid distilling over into the absorption bottles. This delivery tube carries the gas into a series of three bottles, each having a capacity of about 100cc, and containing 30 to 40cc of a $\frac{1}{2}$ per cent. solution of permanganate of potassium.

Add to the flask 2 or 3 grms. of pure Na_2CO_3 which will liberate CO_2 on the addition of the acid and expel the air.

Now run in cautiously and little at a time, 50 to 60cc of conc. HCl. This must be so done as not to cause too rapid an evolution of gas. When the action has become moderate heat carefully to boiling and boil till steam condenses in the tube and all the iron is dissolved, probably 1 or 2 minutes. This will drive all the H_2S over into the permanganate bottles.

Be careful on withdrawing the lamp to let air into the flask by opening the stopcock of the funnel tube, or the vacuum formed on cooling will draw the permanganate back into the flask.

Empty the bottles into a beaker, wash them out, as well as the connecting tubes. Dissolve any MnO_2 adhering to the sides by a little conc. HCl, and add it to the permanganate solution in the beaker. Add in all about 10cc of HCl; now boil and drop in a solution of pure oxalic acid until the MnO_2 all dissolves and a clear solution remains (always look out for traces of unoxidized sulphur in the liquid, which, if found, of course vitiate the analysis). Now add 5cc of a 10 per cent. solution of $BaCl_2$. Let the precipitate of $BaSO_4$ settle. Filter, wash, ignite and weigh.

Examination of the Residue in the Flask for Sulphur.—Filter the contents of the flask on a 9 c. m. filter. Wash the residue thoroughly and dry it rapidly, but avoid heating it above 100°C. Open out the filter and brush the residue into a small beaker. Add 10cc of concentrated HNO_3 and a small crystal of $KClO_3$. Boil down to dryness, take up by heating with 5cc of conc. HCl, dilute to 30 or 40cc, filter and add .5cc of $BaCl_2$ solution to the filtrate. Let stand till any precipitate of $BaSO_4$ settles, filter it off and weigh it.

The residue may also be treated by fusion exactly as an iron ore, of course using smaller quantities of fluxes.

Drown — Trans. Am. Inst. Min. Engrs., vol. II, p. 224.

Many other absorbents are used, for details see the following:

Trans. Am. Inst. Min Engrs., vol. X, p. 187. Cadmium salts.
Trans. Am. Inst. Min. Engrs., vol. XII, p. 507. Bromine.
Blair — Chem. An. Iron and Steel, p. 59, et. seq., Lead and Silver salts and Peroxide of Hydrogen.

DETERMINATION OF SULPHUR IN IRON AND STEEL BY TITRATION WITH IODINE.

This is a very rapid method. It is in general use and when properly conducted gives satisfactory results.

The gas from the evolution flask is passed into a solution of sodium or potassium hydrate by which the H_2S is rapidly and completely absorbed, alkaline sulphide being formed. ($2 NaHO + H_2S = Na_2S + 2 H_2O$.)

On adding HCl to this, the H_2S is again liberated ($Na_2S + 2 HCl = 2 NaCl + H_2S$.) If this be done in the presence of a *large volume* of cold water, the H_2S will not escape, but will remain in solution in the water. This solution of H_2S is then titrated by a standard solution of iodine, which unites with the hydrogen, setting free the sulphur, ($H_2S + 2 I = 2 HI + S$.)

By adding a little solution of starch to the liquid, the least excess of iodine is shown by an intense *blue* color ("iodide of starch").

The liberated sulphur causes the liquid to become curiously opalescent and show various colors, but this does not at all obscure the "end reaction," which is very sharp.

The solutions needed are standard iodine and starch.

Preparation of the Starch Solution.—Mix about one grm. of starch with 3 or 4cc of cold water till all lumps are gone, pour this slowly into about 200cc of boiling water. Boil the liquid a minute or two, then let stand till cold. Do not use it hot.

Preparation of the Iodine Solution.—Weigh 3.968 grms. of pure resublimed iodine on a watch glass. Put it in a small beaker, add about six grms. of pure potassium iodide (free from iodate) and 10cc of water. Let stand in the *cold* until *all* the iodine dissolves, then transfer to a graduated flask and dilute to one litre.

One cubic centimeter of this solution should be equivalent to 0.0005 grms. of sulphur. If five grms. of metal be taken for the analysis each cc of iodine solution consumed will be equivalent to 0.01 per cent. of sulphur.

In the reaction $H_2S + 2I = 2HI + S$, two atoms of I are equivalent to one atom of S, or 254 by weight of $I = 32$ of S. To find the amount of I, which must be contained in 1cc to give a solution of the above value in sulphur, make the proportion $.0005 : x = 32 : 254$, which gives $x = .003968$, and this amount in 1cc will be 3.968 grms. per litre.

Iodine is insoluble in water, but is easily and rapidly dissolved in *very concentrated* solutions of KI, though very *slowly in dilute* ones.

The iodine solution is not constant; hence, its strength must be determined frequently.

Standardizing of the Iodine Solution.—Prepare the following solutions:

A. 7.75 grms. crystallized sodium hyposulphite dissolved in water and diluted to one litre.

B. 0.1536 grms. of fused potassium bichromate dissolved in water and diluted to 100cc.

Measure with a pipette 10cc of solution A into a beaker, add 100cc of water and 3 or 4cc of starch solution. Now run in the iodine solution from a burette until the last drop gives a decided blue color, not disappearing on stirring. Note exactly the amount used. Repeat this two or three times. The results should agree almost exactly. Take their average as the amount of iodine equivalent to 10cc of the "hypo" solution.

Now measure in the same way 10cc of solution B into a beaker, add 50cc of water and then about 0.5 grms. of pure KI.

When the KI is dissolved add 5cc of conc. HCl, which must be free from Cl or Fe. Let the mixture stand *without warming* 6 or 7 minutes, (it will become brown from the liberated iodine) then add 10cc of solution "A," and 3 or 4cc of starch solution. If the liquid now has a dark blue color add 10cc more of solution A, which will make it colorless.

Finally add the *iodine solution* carefully until the blue color is developed. Note exactly the volume of this solution used. Call it R.

Now deduct from the amount of iodine solution which would be equivalent to the 10 or 20cc of the hyposulphite solution used (solution A), this amount (R) and the difference will be *the volume of the iodine solution which is equivalent to 0.005 of sulphur*. Represent this by Q. Then $\frac{0.005}{Q}$ will be the amount of sulphur to which 1cc of the solution is equivalent. To reduce any observed volume to the true value make the proportion $Q : 10 = N : x$, in which x is the true volume of correct iodine solution.

The reactions upon which this process of standardizing depend are:
1. $Na_2S_2O_3 \; 5 \; H_2O + 2 \; I = 2 \; NaI + Na_2S_4O_6 + 5 \; H_2O$. (The crystallized hyposulphite contains 5 molecules of water.)
2. $K_2Cr_2O_7 + 6 \; KI + 14 \; HCl = 8 \; KCl + Cr_2Cl_6 + 7 \; H_2O + 6 \; I$.

The molecule of hyposulphite weighs 496, and as 1 molecule "Hypo." is equivalent to 2 atoms of iodine (254) we have the proportion $496 : 254 = 7.75 : 3.968$, 7.75 being the amount of "Hypo." dissolved in 1 litre to give a solution equivalent to the iodine.

The "Hypo." solution is not very constant, hence cannot be used as an absolute check on the iodine solution, but only as a means of comparing it with an absolutely known amount of iodine. This definite amount of iodine is obtained from the action of bichromate on an excess of KI in the presence of HCl. The reaction between KI and $K_2Cr_2O_7$ is not instantaneous, but rapidly becomes complete.

The relation between the $K_2Cr_2O_7$ and the I is $294.5 : 761.1$, hence $0.01536 \; K_2Cr_2O_7$ will liberate 0.03968 grms. of iodine, the amount which should be present in 10cc of the iodine solution of its strength were exactly right.

The operation itself consists therefore in finding how much of the iodine solution to be standardized, is required to titrate the amount of the hypo. solution which is equivalent to exactly 0.03968grms. of iodine, as liberated by the bichromate. Thus we find *first* how much of the iodine solution is equal to a certain amount of the hypo. solution; *second*, how much of the same iodine solution is equal to what is left after the same amount of hypo. has been acted upon by 0.03968 grms. of iodine, and the difference is obviously the amount of iodine solution which contains 0.03968 grms. of iodine, that is to say will be equivalent to 0.005 of S; but this should be 10cc, hence the difference between 10 and the amount taken is the amount the solution is off the standard.

The only precautions to be noted are the necessity of giving *time* for the $K_2Cr_2O_7$ to react on the KI and the necessity of avoiding *heat* as iodine is readily volatilized from its solution.

The HCl used must be free from all impurities which liberate iodine from KI (Cl, Fe_2Cl_6, $CuCl_2$, etc.).

The Iodine Solution may also be Standardized against a sample of iron or steel similar to that to be analyzed in which the sulphur has been accurately and repeatedly determined by the permanganate method.

Put the metal through the regular process and thus determine to just how much sulphur each cc of the iodine solution is equivalent. Make a factor of correction to apply to the iodine solution. This method of standardizing has the advantage of causing all errors of solution, evolution and oxidation to affect the standard and sample alike.

See Wilson, Jour. An. and App. Chem., vol. V, p. 439.

The Process.—Arrange a flask as in Drown's permanganate method, except substitute for the bottles one large U-tube and a large test tube, into which the connection tube from the U-tube passes, reaching nearly to the bottom.

Put into the U-tube and the test tube each 30cc of a 20 per cent. solution of pure sodium or potassium hydrate.

Now weigh into the flask five grms. of the metal drillings. Add conc. HCl and evolve the gas exactly as in the other process, but *leaving out* the Na_2CO_3, and pushing the process more rapidly, as the NaHO will let no H_2S escape, even if the evolution of gas is very rapid.

When the iron is dissolved and the solution has been boiled, detach the U-tube and test tube. Empty and rinse them into a large beaker or dish, and add 300 or 400cc of cold water. Now add enough HCl to make the liquid distinctly acid, then 3 or 4cc of the starch solution and titrate with the iodine solution, adding it until a drop changes the opalescent liquid to a deep blue, not disappearing on standing two or three minutes.

If five grms. of metal were taken, then the number of cc used, after correction for standard will give the amount of sulphur in hundredths of a per cent.

A blank test on the NaHO must always be made, as it will usually consume a little iodine. This must be deducted from that used in the analysis.

A very simple method of avoiding calculation in this process is to take, instead of five grms. of iron, *ten times the amount in grms. to which the standardizing has shown one litre of the iodine solution to be equivalent in sulphur.* If this is done the number of cubic centimeters used will always be the per cent. of S in hundredths. For example, suppose the test has shown 10.3cc of iodine to be equivalent to .005 of sulphur instead of 10cc, 1cc of this iodine will be equivalent to .0004854 grms. of sulphur and one litre to 0.4854 grms.

Now, if 4.854 grms. be taken for the analysis, each cc of iodine solution will be equivalent to 0.01 per cent. of S.

This method of applying the factors of volumetric solutions to the amount weighed out may be applied to any volumetric process, as for phosphorus, iron, or manganese.

Additional Notes on the Process.—It is essential that everything be carried out *promptly*. The NaHO solution must not be allowed to stand before titration, or the sulphide will oxidize.

The diluted solution, after the addition of HCl, will change very rapidly if not titrated *at once*.

It is desirable that the NaHO be free from iron, which will form ferric chloride and destroy the H_2S. Prepare the alkali solution, let it stand till any precipitate of Fe_2O_3 settles, and use the clear solution.

By adding a drop of phenolpthaline solution to the dilute soda solution, the neutralization by HCl is made easy as the red color of this indicator is at once discharged when the HCl is in excess.

Modifications of the Iodine Method.— The most important consists in substituting an ammoniacal solution of a cadmium salt for the NaHO. The sulphur is absorbed and precipitated as CdS. This is filtered from the excess of alkaline liquid, filter and all are put in a large amount of water, HCl added, which dissolves the CdS and liberates H_2S, this is then titrated as usual. This method avoids the titration in the presence of any hydrocarbons absorbed by the alkali in the ordinary process. These, some claim, may absorb iodine.

See Blair, Chem. Anal. Iron and Steel, p. 71.

The cadmium solution may be made as follows:

75cc H_2O — 75cc NH_4HO 3 grms. $CdCl_2$.

For *use* add 5cc of this to 95cc of water (formula communicated by Mr. Davis).

THE DETERMINATION OF CARBON IN PIG IRON AND STEEL.

The carbon in gray pig iron occurs principally as "free carbon" or graphite. A smaller portion is combined with the iron as "carbide," constituting the so-called "combined carbon."

In white iron and "chilled iron," as also in steel, especially mild or low carbon steel, most of the carbon is combined.

When these metals are dissolved in hot HCl the graphite is left as a black, scaly residue, while the combined carbon mostly passes off with the hydrogen as hydro-carbon gases.

Carbon is always determined by oxidizing it to CO_2, then absorbing the gas in a weighed amount of KHO, or other absorbent. The gas may also be determined by measuring its volume by the processes of gas analysis. In the hands of very skillful operators this would seem a feasible method.

Jour. Soc. Chem. Ind., 9, p. 768; 10, p. 658; also p. 1033.

The carbon is converted to CO_2 when iron is completely burned in oxygeh gas. Unfortunately this simple method cannot be used in most cases, because when iron fragments are heated in oxygen they become coated with a layer of oxide, which protects the interior from further action, and so a portion of the carbon remains unburned, even hours of heating failing to reach it all.

Hence, this method is only applicable where the iron can be reduced to a *very fine powder*, a condition usually practically unattainable.

The first step in the carbon determination is to liberate it from the iron, by dissolving the latter in some solvent, which will leave all the carbon unattacked as a residue.

The carbon thus liberated is filtered out on to ignited asbestos and then burned to CO_2 in oxygen or by chromic acid and sulphuric acid.

Several methods of solution have been used. The essential condition is that no hydrogen gas be liberated, as it will carry off carbon, and no strong oxidizing agent be present, or it will oxidize carbon.

The most rapid solvent for iron which fulfills these conditions is a saturated solution of ammonium or potassium cupric chloride.

The reaction by which the iron is dissolved is

$Fe + 2(NH_4)CuCl_3 = FeCl_2 + Cu_2Cl_2 + 2NH_4Cl$.

The NH_4Cl serves to hold the cuprous salt in solution and greatly hastens the action.

The solution has a tendency to dissolve organic matter, which is liable to be precipitated subsequently with the carbon in the steel. This is especially true of the ammonium salt. It is very difficult to obtain the ammonium compound free from organic matter, derived from the ammonium salts used in its manufacture. The salt should be thoroughly purified by re-crystallization.

A large excess of the solution is required to prevent the separation of metallic copper with the carbon.

The carbon residue retains chlorides very difficult to wash out, and which cause trouble in the subsequent combustion. This is especially true if any metallic copper is left mixed with the carbon, as it forms basic subchlorides. The spongy carbon is best freed from these chlorides by treatment with HCl and then washing.

For very important papers on the carbon determination consult—
Langley Trans. Am. Inst. Min. Engs., vol. XIX, p. 614.
Shimer Jour. An. and App. Chem., vol. V, p. 129.
Blair, Jour. An. and App. Chem., vol. V, p. 122.

Process — Solution of the Metal and Separation of the Carbon.—Prepare a solution of the double chloride of copper and ammonium or potassium. Use the purest crystallized salt obtainable. Dissolve one part in three parts of pure water, free from grease or organic matter. Then add NH_4HO drop by drop until a slight permanent precipitate forms, let settle, decant off the *clear* solution, and filter the turbid portions through ignited asbestos.

The drillings of metal must be free from all *grease* or intermixed particles of wood, straw or paper. They may be separated from the latter by a magnet, from the former by washing them with pure ether and drying. Care in drilling and handling the sample will render this unnecessary.

Weigh out two grms. of pig iron, three grms. of high carbon steel or five grms. of low carbon steel or wrought iron. Put them in a No. 2 or No. 3 beaker, and add at once 50cc of the copper solution for each grm. taken. Stir the solution continuously until the iron is dissolved. The completion of the reaction is easily recognized by the residue becoming light and "flotant." At first more or less copper will separate, but stirring and time will bring it into solution. Now add 5 to 10cc of conc. HCl, and when all the free copper is dissolved, filter on to an asbestos filter. The asbestos must be thoroughly ignited in air before using to remove any carbonaceous matter.

When the liquid has run through, wash out the beaker and transfer all adhering carbon to the filter, using HCl diluted with its own volume of water. Wash the carbon on the filter twice with the acid, letting it run through slowly to give it time to act. Now wash with water until all the HCl is removed and the filtrate does not react with $Ag\,NO_3$.

The filtrate will be dark colored at first, but when diluted with the HCl and water will become light, and then must be carefully examined to see that no particles of carbon have run through the filter.

In the above process double chloride of copper and potassium may be used in place of the ammonium compound. It is said to be more easily obtained pure.

Some chemists add a little HCl to the copper solution *before* adding it to the metal.

Determination of the Carbon by Oxidation with Chromic Acid and Weighing the CO_2 Produced.

Carbon in any form is rapidly and completely converted to CO_2 when heated with chromic acid and an excess of sulphuric acid, chromium sulphate being formed.

The important condition is the strength of the H_2SO_4. If too weak, the boiling point is too low and carbon will escape oxidation. If the acid is too strong, when the mixture is boiled, oxygen will be given off and white fumes of H_2SO_4 will form, which are difficult to arrest in the purifying apparatus, and are liable to cause high results.

The proper strength for the acid in the reaction is between 1.4 and 1.6 sp. gr. This corresponds to from 50 to 70 per cent. of H_2SO_4 in the mixture.

If the carbon retains chlorine or chlorides, chlorochromic acid gas may form, escape with the CO_2 and be absorbed by the KHO, causing false results unless special means be taken for absorbing it.

The H_2SO_4 used must be purified from all organic matter.

Arrangement of the Apparatus.— 1. An Erlenmeyer flask of about 250cc capacity fitted with a 2-hole rubber cork. Into this is inserted a bulb funnel tube having a glass stopcock, and a delivery tube for the gas.

This latter should be of rather large diameter and so inclined that everything condensing in it shall run back into the flask. It is a good plan to have it cooled by a "water jacket" consisting of a larger tube surrounding it, the space between the two being filled with water.

2. Fit a small "guard tube" filled with "soda lime" into the top of the funnel tube.

This is easily made of a test tube drawn out into a narrow neck which is put through a small rubber cork. Put a layer of cotton on the bottom to prevent any soda lime dropping through.

This serves to purify the air drawn into the flask from all CO_2. It must be arranged so as to be easily removed or connected.

3. Connect the delivery tube with the following purifying and absorbing apparatus (or "train") arranged in the order given.

A. A small (50 to 75cc capacity) bottle containing "pyro

solution," made by mixing 0.2 grm. pyrogallic acid with 5 grms. neutral potassium oxalate, adding water enough to make 20cc and then 2 drops of H_2SO_4 which must make the solution distinctly acid.

This will absorb all free chlorine and chlorochromic acid. Its use is due to Langley.

B. A similar bottle containing about 20cc of an acid solution of silver sulphate.

This serves to absorb any HCl vapors. It must follow "A" as the action of the pyro is to form HCl from the oxides of chlorine.

The silver sulphate is easily made by dissolving about 0.5 grm. of $AgNO_3$ in a little water adding 1cc conc. H_2SO_4, evaporating till the HNO_3 is all expelled cooling and diluting largely with water. Ag_2SO_4 is only sparingly soluble.

C. A bottle containing 20 or 30cc of conc. *pure* H_2SO_4.

This takes out all the water vapor from the gas.

D. A U-tube containing granular $CaCl_2$. Fill about an inch of the tube, on the side next to the H_2SO_4 with cotton and moisten the top of this with a drop of *water*. (Blair.)

The object of this $CaCl_2$ is to absorb H_2O and to bring the gas stream entering the absorption apparatus (E and F) into the same condition as to moisture, in which it leaves it. H_2SO_4 will dry air more completely than $CaCl_2$, hence if the gas entered through H_2SO_4 and left through $CaCl_2$ it would carry out more moisture from the KHO bulbs than it brought in and so result in loss of weight. The introduction of the moist cotton is only necessary when the $CaCl_2$ is very dry, after it has absorbed a little water, it will itself give up enough to the air to serve the purpose.

Dried $CaCl_2$ and not the fused salt should be used. This latter is usually alkaline from free CaO and will absorb some CO_2.

E. Liebig's potash bulbs containing a clear solution of KHO of about 1.27 sp. gr., (about 30 per cent.).

This absorbs the CO_2, but not *completely* unless the gas stream is slow. The solution loses water vapor to a small extent. If made stronger than directed it deposits K_2CO_3 and may clog up the tube.

F. A small U-tube, the limb next the potash bulbs filled with granular soda lime which should not be too dry. The other limb is filled with granular $CaCl_2$.

This tube serves to catch any trace of CO_2 escaping the bulbs, and also to retain all moisture. Soda lime is a more rapid and complete absorbent for CO_2 than the bulbs, but it is rapidly exhausted, hence by letting the bulbs do most of the work and only using the soda lime for the finish it lasts for several operations and retains every trace of the CO_2. The potash bulbs and the soda lime — $CaCl_2$ tube are the parts of the train to be *weighed*.

G. A U-tube similar to the the last, but larger, having the limb next to F filled with $CaCl_2$, and the other with granular soda lime.

This serves as a guard tube to prevent moisture or CO_2 working back into the absorption apparatus from the aspirator. It can be used almost indefinitely.

H. An aspirator for sucking air slowly through the apparatus.

This must be easily attached and detached. It can be made from a five-pint acid bottle by boring a hole near the bottom with a pointed file dipped in turpentine, fitting a glass tube in this by a rubber ring and then attaching to this a rubber tube and pinch-cock.

Notes on the above Apparatus.— It is essential that none of the chromic acid solution come in contact with the rubber corks or connections, as it would of course form CO_2. For similar reasons it is necessary that the glass stopcock in the funnel tube be free from grease of any sort.

A flask provided with a ground glass cap, into which the tubes are fused, is a very good substitute for the corked flask described. It is, of course, more expensive.

Liquids always absorb some little CO_2, hence the volume of all absorbing liquids used in purifying must be small. This CO_2 is, however, given up again to a current of air passed through them for some time.

The details of the mechanical arrangement of the train will vary with the operator. For drawings showing convenient plans see

Jour. An. and App. Chem., vol. V., page 336.

Setting Up and Testing the Apparatus.—The connections are made by glass tubes united by short rubber tubes. These must be carefully tied with thread or wire, as it is essential that the whole apparatus be air tight. Rubber corks are, of course, the best, but good ordinary corks can be used if rolled soft and carefully bored and fitted. Sealing wax is often recommended, to make joints tight, but it is a bad thing to use, as it will crack and leak unexpectedly. The potash bulbs and U-tube must be "capped" with short rubber tubes closed with bits of glass rod. These must always be removed for a moment and then replaced just before weighing, that the air pressure inside and outside may equalize itself.

It is necessary to first pass some CO_2 through the apparatus in order to saturate any alkaline material present in the $CaCl_2$, etc. When this is done of course the weighed part of the train is omitted. Connect up the train, leaving out the parts E, F and G. Put a little marble in the flask, add a little dilute H_2SO_4 so as to generate a slow stream of CO_2. Let this run through this portion of the train for about thirty minutes. Disconnect and wash out the flask, replace it, and now aspirate air alone until six or eight litres have been slowly drawn through.

Now, connect up the whole apparatus and attach the aspirator. Close the stopcock in the funnel tube of the flask and see if all connections are *tight*. This is proved by the water ceasing to run from the aspirator. Cautiously let in air by opening the stopcock. Attach the soda lime guard-tube to the funnel tube and aspirate one or two litres of air carefully (not over one or two bubbles a second). Disconnect the bulbs and the U-tube, cap them, and wipe them carefully. Set them in or near the balance case until they attain its temperature (ten or twenty minutes). Uncap them a moment, replace the caps and then carefully weigh them. Replace the apparatus and aspirate three or four litres of air more and reweigh them as before. The KHO bulbs will lose (due to moisture) and the U-tube will gain weight. The loss in one must equal the gain in the other. The total weight of the absorption apparatus must not change more than one-half millegramme.

The Treatment of the Carbon in the Residue from the Iron.—Transfer this to the flask, using a little water to wash out the filter tube. The total amount of liquid in the flask must not exceed 30cc. Now dissolve four grms. of chromic acid in 4cc of water, and pour it into the flask through the funnel tube. Wash out the tube with 2 or 3cc of water. Now estimate the amount of liquid in the flask by comparison with a similar one, and put in the bulb of the funnel a quantity of pure concentrated H_2SO_4, equal to about twice the volume of the liquid in the flask.

This acid should be previously purified by adding to a quantity of it a little chromic acid, heating it to 200°C and letting it cool. This will destroy any trace of organic matter it may contain.

Allow the acid to run into the flask gradually and carefully to avoid too violent action. Shake the flask up carefully to mix the solution. The evolution of CO_2 will begin at once. Finally heat with a lamp until the liquid begins to boil. Continue this till no more bubbles come over through the bottles in the train. A sharp rattling sound usually marks this point. The time of boiling should not exceed one or two minutes. Now withdraw the lamp and immediately open the stopcock of the funnel tube to admit air and prevent back suction. Connect the funnel tube with the soda lime guard tube, and let the apparatus cool a few minutes. Then aspirate carefully four or five litres of air (or more if the apparatus is large). Detach the absorption apparatus and weigh as before. The total gain in weight will be the amount of CO_2, and this multiplied by 0.2727 gives the amount of carbon.

The greatest care and "handiness" are necessary, but with skill duplicates should agree within 0.01 per cent.

Note 1.—The "weighing out" of pig iron for carbon is a matter of great difficulty, as the fine, dusty portion is usually higher in carbon than the lumps. A method proposed by Dr. Shimer is to moisten the drillings with alcohol, so that the fine may stick to the coarse. Then take a portion of approximately the right amount, put it on a weighed watch glass, dry it carefully and reweigh, using this amount for the determination.

Note 2.—This method of combustion with chromic acid, while less elegant than the combustion by oxygen gas in a glass, porcelain or platinum tube is accurate, and demands much less expensive apparatus.

For the combustion in oxygen see—

Blair Chem. Anal. Iron and Steel, p. 125, et seq ; also Langley paper Trans. Am. Inst. Min. Engs., vol. XIX, p. 614.

THE DETERMINATION OF CARBON IN STEEL BY COLOR.

This method is in universal use in steel works. It depends upon the fact that when steel is dissolved in dilute HNO_3 there separates a brown compound containing the carbon, which on boiling goes into solution giving a color which is deeper, as the per cent. of carbon is higher. Pure iron dissolves in HNO_3 giving a nearly colorless solution, from which moderate dilution removes every trace of color.

The color produced by the carbonaceous matter is rapidly altered by light. Its depth depends somewhat on the mode of solution, the concentration of the acid and the kind of steel; hence the process must be conducted strictly according to rule to get good results.

There is required, first, a standard steel, which must be of exactly the same kind as that to be tested, and also be similar in its composition and of approximately the same carbon percentage. The carbon in this must have been accurately determined gravimetrically; second, nitric acid of 1.2 sp. gr. perfectly free from chlorine, as the least trace will seriously alter the color of the iron solution, making it more yellow. The above strength corresponds to 1 of conc. acid to 1 of water; third, comparison tubes of clear white glass, graduated in $\frac{1}{10}$ cc, and of exactly equal diameter.

For comparing the colors a blackened box with a ground glass window in the end is convenient.

See Blair, Chem. An. Iron and Steel, Second Edition, p. 167.

Process. — Weigh 0.2 grm. of the steel and of the standard, each into a 6-in. test tube. Add to each tube a measured volume of cold HNO_3 1.2 sp. gr., using the following amounts: For steels with not over $\frac{2}{10}$ per cent. carbon, 4cc; from $\frac{2}{10}$ to $\frac{5}{10}$ per cent., 6cc; from $\frac{5}{10}$ to 1 per cent., 8cc; and over 1 per cent., 10cc.

Stand the tubes in cold water till violent action ceases. Then set them in a water bath kept boiling and boil until the solution is perfectly clear and no more bubbles of fine gas appear. Keep the mouths of the test tubes closed loosely by little glass bulbs or balls to prevent drying of the iron salts on the sides of the tubes. The time required will be 15 to 30 minutes, according to the carbon contents of the steel. Now cool the tubes in water. Add an equal volume of water to each and pour into the comparison tubes. Dilute carefully until the colors match. The percentages of the carbon will be to each other as the volumes in the tubes.

Where a number of steels are to be tested at once it is convenient to dilute the standard until each cc represents some definite percentage of carbon, and then match it with the others, so that the readings in cc's can be at once converted to per cents. For example, if the standard contained 0.38% C, dilute it to 19cc, then if a comparison showed the unknown steel to read 16cc it would obviously contain 0.32% carbon.

Note.—If the steel contains much sulphur the solution will be slightly *turbid* from free S. Comparison is difficult in this case.

Many modifications of this process are proposed. First, by the use of permanent standard colors to avoid dissolving a standard steel every time. Both organic and inorganic colors (chlorides of Fe, Cu and Co), are used, but the safest way is to do as above described.

For these methods see Am. Inst. Min. Engrs., vol. XVI, p 111, also vol. I, p. 240.

For very low carbon steels the color is very faint and uncertain. In such cases an alkaline method has been used.

Stead, Jour. Iron and Steel Inst., 1883, No. 1, p. 213, or Blair, Chem. Anal. Iron and Steel, Second Edition, p. 170.

DETERMINATION OF "GRAPHITE" OR UNCOMBINED CARBON IN PIG IRON.

Treat 2 grms. in a beaker with 50cc of HCl, sp. gr. 1.12. Cover and boil briskly for 30 minutes. Dilute, filter on an asbestos filter and wash first with hot water, then with a solution of caustic soda, then with water, then with alcohol, then with ether and finally with water first cold then hot, till every trace of ether is extracted. Now transfer to the carbon apparatus and treat with chromic acid and sulphuric acid as in the determination of total carbon.

Drown, Trans. Am. Inst. Min. Engrs., vol. III, p. 41.

This complicated washing is required to remove solid and liquid hydrocarbons which are liable to form and are insoluble in water alone.

DETERMINATION OF TITANIUM IN IRON ORES.

Titanium usually occurs in iron ores as menaccanite. It is seldom found except in magnetites.

If titaniferous iron ores are boiled with concentrated HCl most of the TiO_2 goes into solution, provided the material is reduced to an excessively fine powder.

Rutile, ignited TiO_2 and other compounds of titanium, which are not attacked by HCl, can be dissolved after fusion with acid potassium sulphate ($KHSO_4$). The fusion is slowly but completely dissolved by cold water. Titanic acid and iron, with other bases going into solution as sulphates, while only silica is left as a residue. Should phosphoric acid be present, some TiO_2 is liable to remain as phosphate with the SiO_2 however.

When ores or other compounds containing titanium are fused with a large excess of dry sodium carbonate, they are completely decomposed, provided they are very finely pulverized. When the "melt" is boiled with water till thoroughly disintegrated, sodium phosphate, sodium aluminate and sodium silicate go into solution, while sodium titanate remain entirely insoluble with the oxide of iron and other bases. The separation in the case of the phosphates is complete, in the case of silicate and aluminate partial.

The residue of titanate, etc., is now soluble in HCl or even more readily in hot H_2SO_4. The solution may be freed from silica by evaporation with H_2SO_4 till HCl is expelled and the silica dehydrated. This will also dissolve any TiO_2 the HCl does not attack. On dilution with water the TiO_2 and the bases will all go into solution as sulphates, while the SiO_2 is left insoluble and free from TiO_2.

TiO_2 is precipitated from slightly acid solutions on boiling — as a hydrate. That the precipitation may be complete, it is necessary that not more than one-half per cent. of free acid be present, that the solution be dilute, the boiling prolonged and that phosphoric acid be absent or present only in traces. Ferric oxide, alumina and phosphoric acid, if present, invariably come down with the TiO_2 in considerable amount.

The precipitate tends to adhere to the glass, is fine and difficult to filter and wash.

The precipitated titanic hydrate is soluble with difficulty in mineral acids, and is insoluble in acetic acid.

Solutions containing TiO_2 are completely precipitated by ammonia. TiO_2 is also precipitated on boiling completely and promptly from solutions containing sodium acetate and a large excess of acetic acid (15 to 20 per cent.), provided *ferric salts* are not present. If phosphoric acid is present it will come down with the TiO_2. As alumina is not precipitated under these conditions, the TiO_2 will be nearly free from it, and a repetition of the process will give a good separation.

<small>See Gooch. Chem. News, vol. LII, p. 55.</small>

The best solvent for hydrated TiO_2 is a mixture of eight parts of conc. H_2SO_4 and three parts of water, heated to its boiling point.

If this solution is concentrated by boiling until the water is all expelled, the TiO_2 will separate again in a form which cannot be redissolved; hence, in the process which follows *avoid too high heating of the sulphuric acid solutions.*

Process for Iron Ores.— Take one or two grms. (the ore must be ground to an impalpable powder), put it into a small covered beaker, add 30cc of concentrated HCl, and boil gently till the iron appears dissolved. Now add 10cc to 15cc of dilute H_2SO_4 (1 to 4 of water), boil down till the HCl is all expelled and fumes of H_2SO_4 just begin to appear, but avoid over-heating. Cool; add 25cc of H_2O and boil till all iron salts are dissolved; filter and wash. Dry and save the *residue* to be examined for TiO_2. It is usually free from it, but may contain a considerable amount.

Dilute the filtrate to 250cc or 300cc. Add NH_4HO carefully until the precipitate formed at first only dissolves slowly on stirring. Now warm the solution and add slowly, a little at a time, a solution of sodium sulphite, 1 in 5, made distinctly acid by adding a little H_2SO_4. The deep color produced by each addition will rapidly disappear as the iron is reduced.

Should the liquid grow turbid and a precipitate form, add a few drops of HCl to clear it, and continue with the sulphite solution, giving time for the reaction. The solution should not be heated to boiling or TiO_2 will separate as a white, milky precipitate, not redissolving by the addition of a few drops of HCl. This will do no harm except as it makes the reduction difficult to follow. Add in this way 40cc to 60cc of the sulphite solution. The liquid should now be colorless and smell strongly of SO_2. If still colored from ferric iron a few moments' heating will reduce it. Now add 50cc to 60cc of acetic acid, and then 20 grms. of sodic acetate, and boil hard for three minutes. The TiO_2 will separate as a flocculent precipitate. Let this settle, filter, and wash with hot water. Call this precipitate A; it contains all the TiO_2 but is impure.

Treatment of the Residue.— Put filter and residue into a platinum crucible and ignite until all carbon is burned off. Now add ten times its weight of dry Na_2CO_3 and fuse over a blast lamp until thoroughly decomposed. Cool, add water and boil until all is thoroughly disintegrated and the residue is flocculent. Filter and wash with hot water. Wash the residue from the filter into a beaker with a little water. Let it settle and decant the water off through the filter again. Dissolve the little adhering residue from the filter with a few drops of HCl, running it through into the rest. Add a little HCl to the beaker and residue, then 10cc of dilute H_2SO_4 (1 to 4) and boil; all will dissolve. Evaporate until all HCl is expelled, just as with the original solution. Add 25cc H_2O, boil and filter from the SiO_2, which will be free from TiO_2. Now dilute to 150cc, add NH_4HO till a slight precipitate forms; redissolve this with a little HCl, and add 10cc of the sulphite solution. If a precipitate now forms, add HCl till it just dissolves. Now heat gradually; as soon as the solution is colorless add 30cc of acetic acid and 10 grms. of sodium acetate. Boil three minutes. If any TiO_2 separates, filter it off and wash as before. Call this Precipitate B.

Treatment of the Impure Precipitate.— Put Precipitates A and B into a platinum crucible. Burn off the filter papers completely, and weigh the impure TiO_2. Add ten times this weight of dry Na_2CO_3 and fuse thoroughly. Boil out the fusion with water till completely disintegrated; filter and wash. Now wash the residue off the filter into a beaker; let the liquid settle and decant the clear liquid back through the filter (the object of this is to get the residue into the beaker, yet have very little water present).

Treat the filter paper with a little conc. HCl, letting it run through into the residue in the beaker (this dissolves most all the TiO_2 adhering to the paper). Finally burn the paper at as low a heat as possible and add the ash (still retaining a trace of TiO_2) to the beaker. Dissolve the contents of the beaker in conc. HCl, then add 10cc or 15cc of dilute H_2SO_4 and evaporate as before. Dilute with 25cc of water, boil, filter and wash. Dilute the solution to 250cc and precipitate the TiO_2 exactly as in the original solution, except that only 20cc of sodium sulphite will be required for the reduction. The second precipitate of TiO_2 should be white and pure. It is dried, ignited intensely and weighed.

The ignition should be repeated and the precipitate again weighed until its weight is constant. The TiO_2 must be light colored.

The use of H_2SO_4 in the above process serves two purposes. It separates SiO_2, and when concentrated to the right strength it is the most powerful solvent for TiO_2, dissolving even the ignited oxide provided too intense ignition has been avoided.

PROCESS FOR ORES NOT ATTACKED BY HCl.

Prepare pure bisulphate of potash by melting it in a platinum dish or crucible until all boiling ceases and it is in quiet fusion. Cool and pulverize it and keep in a tight bottle.

Mix the very finely pulverized ore with 15 times its weight of the bisulphate. Transfer to a large platinum crucible and heat gently till the mass melts. Raise the heat gradually until white fumes just begin to come off (a low red) and keep in quiet fusion 20 or 30 minutes. Avoid too high a heat or the H_2SO_4 will be driven off and the fusion spoiled. Insert a platinum wire into the melted mass and let the crucible cool. The fusion will generally detach itself from the crucible and can be lifted out by the wire. Now put crucible and cover into a beaker. Add a considerable amount of cold water. Hang the lump of fused material in the water and let it soak out. Everything will gradually dissolve except SiO_2, though it may take 12 hours. Now filter from the residue of SiO_2 and proceed as with the ordinary solution.

The bisulphate fusion may be made to dissolve more rapidly, if after cooling in the crucible there is added a little conc. H_2SO_4, and the whole again heated till it melts. On again cooling the mass will be pasty and dissolve more rapidly. (Kennedy).

See Drown, Trans. Am. Inst. Min. Engrs., vol. X, p. 137.
Jennings, Eng. and Min. Jour., vol. XLV, p. 475.

DETERMINATION OF TITANIUM BY COLOR.

When hydrogen peroxide (H_2O_2) is added to a sulphuric acid solution of titanium, a yellow brown coloration is produced, the intensity of which is proportional to the amount of TiO_2. This is a very delicate qualitative test and has been made the basis of a color method. The results are satisfactory for small percentages, as the solution of TiO_2 must be very dilute in order to compare colors well. For details see

Noyes, Jour. An. and App. Chem., vol. V, p. 39.
Wells, Trans. Am. Inst. Min. Engrs., vol. XIV, p. 763.

THE ANALYSIS OF COAL AND COKE.

The proximate analysis is that usually made. This gives information as to the products of decomposition, but is of little value as determining the heating power of a coal.

It is necessary to conform strictly to the method of analysis that the results may be uniform, as small variations in the process give large differences in results.

Coals rapidly lose moisture and change in other respects when kept in the powdered condition, hence the analysis must be made without delay after sampling.

When a sample of coal is dried at 100° or 110°C, it will lose weight for a while and then gain from oxidation, hence weighing after a definite time must be substituted for drying to a constant weight.

Process for the Proximate Analysis. —Weigh 1 grm. of the coal into a platinum crucible provided with a well fitting lid. Put it uncovered in a drying oven and keep at 105°C for *one hour*, cool and weigh. The loss in weight is called "*moisture*," and represents with sufficient exactness the hygroscopic water in the coal.

Now cover the crucible and heat it over a Bunsen burner for *three and one-half minutes*. Then, without allowing it to cool, substitute a blast lamp for the Bunsen burner and heat three and one-half minutes longer. Now cool it rapidly and weigh. The loss in weight is called the "*volatile combustible matter*." The crucible is now set over a burner, and so inclined that the air can circulate in it, setting the cover against the end to help the draught. The carbon gradually burns out, and, when the ash appears free from it, the crucible is cooled and weighed. It is again set over the lamp and burned a few minutes longer and again weighed. When the weight becomes constant, what is left is called "*ash*," and the loss in weight called "*fixed carbon*."

Of course the sum of the "moisture," "volatile combustible matter," "fixed carbon" and "ash" will always be 100 per cent. The sum of the fixed carbon and the ash gives the "*coke*" produced.

The above method is that which has been used for all the analyses of the Ohio Geological Survey since 1880. It yields fairly constant results. The only important variation in use consists of taking the coal *without previous drying*, for the determination of the volatile combustible matter and then determining the *moisture* in a separate portion. This gives 1 to 2 per cent. less fixed carbon and more volatile matter, but is a little more trouble, and as the results are simply those yielded by a conventional method, the above is recommended.

See Cairn's Quant. Anal., p. 238; Hinrichs, Chem. News., vol. XVIII, p. 53.

Determination of Sulphur in Coal—Eschka's Method.—
Sulphur exists in coal in three forms: Pyrites, "organic sulphur" and sulphates. By heating coal with a mixture of MgO and Na_2CO_3 and ample access of air, all unoxidized sulphur is converted to sulphites and sulphates of soda and magnesia. On boiling out the burned mass with water these, as well as any sulphuric acid existing previously in the coal as sulphate, are all dissolved out as alkaline salts. By adding bromine water to the solution all sulphites are oxidized to sulphates, and thus the total sulphur can be estimated as $BaSO_4$ by precipitation with $BaCl_2$.

*Preparation of the Soda Magnesia Mixture (Eschka Mixture).—*Take a good quality of commercial "light calcined magnesia" and purify it from sulphur as follows: Add about two per cent. of C. P. sodic carbonate, and then stir it up in enough boiling water to make a thin liquid. Boil the mixture a few minutes and let settle, decant off the liquid by a siphon. Add water again, stir up, settle, and again decant. Continue this washing by decantation until the liquid after being acidified with HCl shows no trace of sulphates when tested with $BaCl_2$. Now pour the MgO on to a large filter, let it drain and dry. It is now free from sulphur compounds.

Ignite the dry MgO in a covered platinum crucible until all water is expelled. Cool and weigh it. Now add half its weight of previously dried C. P. Na_2CO_3. Grind the two together till thoroughly mixed, and keep in a tight glass stopped bottle.

Of course if a sample of light oxide of magnesia can be obtained free from sulphur the above is unnecessary. But by using a tin bucket to work in the preparation from ordinary material is easy and the results sure.

*Process of Analysis.—*Weigh one grm. of the coal or coke (which must be finely powdered, especially in the case of coke), then weigh out roughly two grms. of the

"Eschka mixture." Mix about two-thirds of this with the coal, using a spatula and a piece of glazed paper. Transfer this to a 30cc platinum crucible, and settle it down by tapping the crucible on the table. Now cover the contents of the crucible with the remaining portion of the Eschka mixture.

Set the crucible in an inclined position, over a small *alcohol* flame, so that the tip of the flame may barely touch the crucible near the *top of the mixture*. The heat must be carefully regulated, so that no blackening of the white cover layer takes place, and no trace of smoke appears. The cover should be set in the mouth of the crucible to assist the draught. The mixture soon ignites and will gradually burn through, as may be observed through fissures which open in the mass. The coal will usually burn completely in less than an hour. The heat may be raised toward the end of the combustion and the lamp set back under the bottom of the crucible. A higher heat may be used with cokes from the start, as these give no volatile products and burn slowly. Finally stir up the powder with a hot glass rod or platinum wire. If the burning is complete all trace of the black coal will have disappeared and only a light, reddish gray mass remain. Cool and pour the powder into a 200cc beaker. Add about 100cc of boiling water, stir and digest on a water bath. Then filter and wash the residue thoroughly with hot water.

To the filtrate add bromine water until the liquid is colored yellow, then add 2 or 3cc of HCl and warm until all CO_2 is removed.

Then add a slight excess of $BaCl_2$ solution, and let the $BaSO_4$ settle. Filter, wash, dry, ignite and weigh as $BaSO_4$. Calculate the sulphur as S.

Always examine the *residue* which was extracted with water, by washing it off the filter into a beaker, and then adding a little HCl and warming. All will dissolve but a little ash. If any *unburned* coal is seen in the residue, the analysis must be repeated.

Note.—The above is by far the neatest and quickest method for sulphur determination in coals. If care be taken in all details, especially as to rate of heating, there is no loss of sulphur whatever.

It has been proposed to substitute K_2CO_3 for the Na_2CO_3, and said that there is less danger of loss of S with the potassium carbonate, but the writer's experience, extending over hundreds of analyses, has shown it to be absolutely unnecessary.

The use of *alcohol* instead of gas as a source of heat is *essential*. All coal gas contains sufficient sulphur to vitiate the results.

A careful "blank" must be run on the chemicals, and any sulphur they contain deducted from that found as above.

Eschka, Zeitschrift An. Chem., vol. XIII, p. 844.
Drown, Trans. Am. Inst. Min. Engs,. vol. IX, p. 660.

THE ULTIMATE ANALYSIS OF COAL.

Determination of the Carbon and Hydrogen by Combustion in Oxygen.—

The coal, placed in a boat of porcelain or platinum, is burned in a combustion tube, through which a current of purified air and oxygen gas is passed. The H_2O and CO_2 produced are absorbed and weighed.

The only difficulty arises from the presence of sulphur in the coal, which forming SO_2 would pass over into the potash bulbs and vitiate the carbon results. This is prevented by the use of lead chromate which oxidizes and holds the S as $PbSO_4$.

To work most effectively the $PbCrO_4$ must be kept at a barely visible red heat, and on no account be allowed to melt.

The process is difficult only in that it requires close attention to details and skill in fitting and manipulating apparatus.

The *chemical* points of difficulty are as follows: First, coal begins to decompose at a low temperature, among the products is marsh gas (CH_4). This is very difficult to burn, and easily escapes from the combustion tube. To secure its oxidation a *long* and *hot* layer of copper oxide is necessary. Second, the oxygen and air passed through the apparatus may easily carry in hydrocarbon vapors taken up from rubber connections, these becoming oxidized in the combustion tube lead to false results. For this reason the oxygen must not be supplied from rubber gas bags, and long rubber tubes are to be avoided. Third, the oxide of copper and asbestos used in the tube must be free from $CaCO_3$, or other carbonates or alkaline bases. Commercial CuO frequently contains these impurities. These carbonates give up CO_2 in the tube on heating, and the CaO left will absorb CO_2 in subsequent tests. Examine the CuO by washing with a little cold dilute HNO_3, then, after neutralizing with NH_4HO, testing the liquid with ammonium oxalate.

The asbestos used in the tube must be first freed from carbonates and organic matter by boiling with hydrochloric acid, washing with water until free from acid and then igniting thoroughly in air.

The best way to secure a pure oxide of copper is that proposed by Blair. Oxidize fine copper gauze by heating in a current of pure oxygen.

All the points apply here that were mentioned in connection with the determination of carbon, as to handling and arranging the absorption apparatus.

The Apparatus and its Arrangement.—First: a glass or metal gas holder containing oxygen.

This may be cheaply made of two 5-pint acid bottles. Bore holes near the bottom and connect the two by a rubber tube. One bottle holds water, the other, provided with a cork and exit tube, contains the gas.

About 500cc of oxygen will be required for a combustion. The gas is best made by heating a mixture of $KClO_3$ with one-third its weight of MnO_2 in a small round-bottomed glass flask.

Second: Purifying Apparatus for the Air and Oxygen, consisting of: A.—A wash bottle with a three-hole cork and containing 50 or 75cc of a clear solution of KHO of 1.27 sp. gr. This is fitted with three tubes. One for oxygen, connected with the gas holder. This connection must be of glass tubes and made flexible by one or two short rubber joints on one of which is put a pinch-cock for regulating the gas supply. A second tube allows air to enter. Both these tubes dip deep into the KHO solution. A third tube conveys the gas and air from the bottle to the next point. B.—A U-tube filled with "soda lime." C.—A U-tube filled with $CaCl_2$.

This apparatus frees the air and oxygen from all traces of CO_2 and moisture. It may be made much more elaborate where it is to be used repeatedly, as this simple arrangement will be soon exhausted.

Third: A Glass Combustion Tube set in a packing of asbestos, in the trough of a long gas combustion furnace.

This must be of the best infusible glass. It should have an internal diameter of about one-half inch, and the glass must not be too thick or it will crack.

The ends are to be rounded by heat and stopped with good soft corks well rolled. Rubber connections with this tube are not to be recommended as they become warm and are liable to give off hydrocarbon vapors.

This tube is filled as follows, leaving a space of two or three inches empty at each end: 1.—A plug of asbestos. 2.—Six inches of coarsely powdered fused $PbCrO_4$. 3.—

Asbestos.   4.—Eight inches of granular, pure, recently ignited CuO (or a close coil of fine copper gauze thoroughly oxidized by heating in a stream of pure oxygen).   5.—Asbestos.   6.—The "boat" for holding the coal.   7.—A coil of Cu gauze two or three inches long, made so that it can be easily withdrawn and replaced. This must be thoroughly oxidized before using.

The end of this tube containing the movable coil and the boat are connected by a glass tube and short rubber joint with the "purifying train" above.

The cork connections in the ends of this tube must not become hot so as to run any risk of burning, hence a sufficient length of empty tube must remain at each end.

Fourth: The Absorbing Train following the combustion tube consisting of: A.—A $CaCl_2$ tube, the end being inserted into the cork of the combustion tube. B.—Leibig's potash bulbs. C.—A soda lime—$CaCl_2$ tube similar to that used in connection with the potash bulbs in the carbon determination.   D.—A guard tube and aspirator.

Testing the Apparatus.—*First*, see that it is *perfectly tight* by running the aspirator and shutting off the entrance of air.

Second.—Heat the tube redhot throughout and aspirate two litres. Detach and weigh the potash bulbs and U-tube. Connect up again and aspirate one-half litre of oxygen and then two litres of air. Detach the tubes and weigh. There should be neither gain nor loss of weight. When these tests are found satisfactory the analysis may be proceeded with.

Process of Analysis.—Ignite and cool the boat. Weigh into it 0.2 grm. of the finely pulverized and mixed coal. (The sample must be made *very fine*, or weighing so small a quantity will not give average results.)   Insert the boat into its proper place. Replace the coil of CuO that follows it. (This should have been only exposed to the air for a moment, as it readily absorbs moisture.)

Connect up the apparatus and then carefully heat the $PbCrO_4$ to *dull* redness and the CuO to bright redness, drawing a slow current of air through the apparatus all the time. Now begin to heat the coal cautiously and introduce oxygen, regulating the supply so as to avoid too vigorous combustion and consequent fusion of the ash, which will lead to retention of carbon. When all is burned cut off the oxygen and aspirate air. Turn off the gas burners and let it cool, continuing to aspirate air until two or three litres or more (at least seven times the capacity of the whole apparatus) is drawn through. Now detach the absorption train and weigh. The increase of the $CaCl_2$ tubes gives the water produced, which divided by nine gives the hydrogen in the coal. The increase in the CO_2 apparatus the CO_2, $\frac{3}{11}$ of which is carbon. The apparatus is now ready for another determination, as the CuO will have all been reoxidized.

THE DETERMINATION OF NITROGEN IN COAL.

This is best done by the Kjeldahl method or by the soda lime method, *either of which* yields correct results, *provided* the coal be in very fine powder and be *completely oxidized*. In the Kjeldahl method the sulphuric acid solution of the coal must be colorless and free from specks of carbon. The time required to accomplish this may be two or three hours, but it can always be done and is essential.

In using the *soda lime method* the soda lime in the tube must be heated until it becomes white and all carbon has disappeared, so that on testing by solution in HCl it shows no remaining carbon.

Fault has been found with the soda lime method for nitrogen in coals; but with these precautions parallel determinations have shown me that it gives the same results as the Kjeldahl and the so-called "absolute" method of Dumas.

For detailed and accurate description of both methods see
Bulletin of the U. S. Dept. of Agriculture, No. 31 or 35.

(Report of the Proceedings of the Association of Official Agricultural Chemists.)

The Oxygen in Coal.—This is estimated by difference, subtracting the sum of the H. N. C. S. and ash from 100. The results so obtained are open to criticism, but no good direct method is known.

THE ANALYSIS OF FURNACE AND FLUE GAS.

The principal ingredients are CO_2, CO, O and N. In addition to these there are small percentages of H, CH_4, C_2H_4, H_2S and occasionally other gases.

The determination of the CO_2, CO and O is all that is usually re-

quired for metallurgical purposes, the residue being considered as nitrogen.

For complete information on the subject of gas analysis the student is referred to

"Technical Gas Analysis," by Clemens Winkler. Translated by Lunge.

"Gas Analysis," by Hempel.

The process for the determination of the three gases named consists in treating a measured volume of the gas with a series of reagents which absorb them successively, and measuring the remaining volume each time.

There are many forms of apparatus used for this operation. That the results may be accurate it is desirable that the tube in which the gas is measured be surrounded by a water jacket. This serves to keep the temperature nearly uniform. By placing a thermometer in the water, any variations in temperature during the course of the analysis may be noted and allowed for, by correcting the corresponding reading so as to bring it to what it would have been, had the temperature remained that at which the original volume was measured. This can be done by using the formula $V = V' - V' \frac{(t' - t)}{273 + t'}$, in which V is the volume at the temperature t, and V' the volume at the temperature t'.

To avoid calculation the gas is always measured at the atmospheric pressure, to which it is brought by a pressure tube or bottle connected with the reservoir by a flexible tube, so the level of the liquid outside and in may be made the same.

Variations of the barometer are not considered, as they are rarely important during the short time the analysis covers.

Collecting the Gas for Analysis.—Take two quart bottles, bore a hole in each near the bottom and connect them by a rubber tube. Fit one with a rubber cork and glass tube, to which attach a small rubber tube closed by a good pinch-cock. Put into the bottles a saturated solution of salt (NaCl), so that by elevating the open one the other can be completely filled (rubber tube and all). Now attach the rubber tube to an iron tube, reaching well into the gas main. Lower the empty bottle, open the pinch-cock and let the liquid slowly run out of the full bottle, drawing the gas after it. When it is full close the cock and disconnect from the gas tube.

Furnace and all gases rich in CO_2 are rapidly changed by standing in contact with water, in which the CO_2 is quite soluble. Salt both diminishes this solubility and makes the absorption much slower. Such gases should be analyzed as soon as possible after the sample is drawn.

Troilius Notes on Chem. of Iron, p. 76.

Samples cannot be kept unaltered in rubber bags.

Determination of CO_2, CO, and O, with A. H. Elliot's Apparatus.—

This is one of the simplest and most easily handled of the various forms of absorption apparatus. It consists of a graduated measuring tube connected by a capillary tube and stopcock, with a plain tube in which the gas can be treated with the various absorbents.

It is figured in the catalogues of most dealers in chemical glassware.

It should be provided with the water jacket and thermometer.

For full description see

Chemical News, vol. XLVIII, p. 189; also School of Mines Quarterly, Nov., 1881.

The Reagents used for Absorption are as follows: They must be applied *in the order given*, as some will absorb other gases besides the special one intended, unless such were first removed.

1. *Absorbent for CO_2.*—A 16 per cent. solution of caustic potash (KHO). This acts very *rapidly* and *completely*. It also absorbs H_2S and SO_2, if present in the gas.

2. *Absorbent for Oxygen.*—An alkaline solution of pyrogallic acid. Dissolve twenty grms. of pyrogallic acid in 100cc of water. This solution keeps fairly well if in tight bottles. When about to use it mix some of this solution with its own volume of the potash solution.

This mixture absorbs oxygen very rapidly at first, but only takes out the last trace after somewhat prolonged contact with the gas. Hence after absorbing most of the oxygen, add a little fresh solution, and let it act ten or fifteen minutes longer.

3. *Absorbent for Carbonic Oxide.*—A strongly acid solution of *cuprous chloride*. Dissolve fifteen grms. of "red oxide of copper" (Cu_2O) in 100cc of strong HCl (sp. gr. 1.19). Keep the solution in a glass stoppered bottle, with scraps of metallic copper.

The pure solution is colorless. Oxygen turns it dark brown. This solution absorbs CO quite rapidly, but only completely when in *very large excess*. The best way to use the reagent is to add successive portions of the fresh solution, so the last used will have only the residual

CO to absorb. This reagent must be applied after the O is removed, as it absorbs it also.

When mixed with *water* this solution deposits Cu_2Cl_2 as a sandy, insoluble white powder, which can only be dissolved out of the apparatus by HCl.

The following Absorbents are also occasionally used: Solution of *silver nitrate* absorbs H_2S only. Solution of *iodine* in iodide of potassium and water (about 3 per cent. of iodine) absorbs H_2S and SO_2. *Bromine water* absorbs ethylene (and others of the series); also H_2S and SO_2. When this reagent has been used, the bromine vapor must be removed (before the gas is measured) by potash solution.

Process for Analysis.—Transfer the gas to the measuring tube. Raise the pressure bottle until the liquid in it and in the tube are at a level. Now read the volume. Do this at intervals of a minute until the readings do not change, read and record the thermometer. Now transfer the solution to the absorbing tube and let the liquid absorbent run slowly down the sides of the tube until no further contraction is observed on standing three or four minutes. Add water to the funnel of the absorption tube and let it run through until the reagent is washed down. Do this gently, avoiding all "splashing." Now transfer the gas back to the measuring tube and measure as before. Record the thermometer each time and reduce all readings to the initial temperature by the formula given.

When the gas is in the measuring tube and the stopcock connecting it with the absorption tube closed, empty the latter and wash it out. Now fill it with water, transfer the gas, and proceed with the next reagent in exactly the same way. Finally, calculate the successive losses of volume into percentages of the original volume.

Extreme care should be taken to avoid getting any of the absorption solutions into the measuring tube. Should this happen the whole apparatus must be carefully washed out before starting a new analysis, as the action on the CO_2 especially will cause loss.

THE ANALYSIS OF BLAST FURNACE SLAGS.

Most slags can be dissolved by hydrochloric acid, especially if they have been suddenly cooled from the melted state.

The few slags which are not attacked by acids must be decomposed by fusion with carbonate of soda. (See analysis of fire clays.)

It is frequently necessary to know the silica, alumina, lime and magnesia in a slag. The following process is rapid and will give sufficiently accurate results for furnace control.

Determination of CaO and MgO.—Weigh one grm. of the finely pulverized slag. Add to it 30cc of water. Stir it well up in the water, so that there may be no "caking" on addition of acid. This is important, for, if the acid be added to the slag while in a compact mass on the bottom of the dish, it is at once covered with a coat of gelatinous silica, which prevents solution.

Now add 20cc of HCl and heat until all is dissolved, except a few flakes of SiO_2 and carbon or sulphur, and no gritty residue remains. Cover and boil to dryness. Heat carefully till the HCl is expelled. Then add 10cc of HCl and 50cc of water. Boil, transfer without filtering to a 500cc flask and add 10cc more HCl. Dilute till the flask is two-thirds full and add NH_4HO till the alumina is precipitated, but avoid a *large excess*. Dilute the flask to the mark, mix well and let settle. Filter off 200cc through a dry filter.

In this, precipitate the CaO as oxalate and then the MgO as phosphate exactly as in the filtrate from the Al_2O_3, Fe_2O_3 in the analysis of limestones. Calculate the results on 0.4 grms. taken.

The precipitate of calcium oxalate may be estimated volumetrically with potassium permanganate. See

Fresenius Quant. Analysis Determination of CaO.

Determination of SiO_2 and Al_2O_3.—Weigh another portion of 0.5 grm. Treat it with water and HCl as before. Dry and add HCl and water, boil, filter and wash. Ignite and weigh the residue, which may be taken as *silica*.

It may be tested by adding H_2SO_4 and HFl, and after driving off the SiO_2, igniting and weighing the residue, which should be deducted from the total weight. (Blair Chem. An., Iron.) As, however, the silica is never completely separated by a single evaporation, the impurities present will about balance the silica lost, so that for ordinary furnace control the gross weight will be reasonably accurate.

To the *filtrate* from the silica add 20cc of HCl, then a slight excess of ammonia. The solution should be diluted to about 200cc, warmed but *not* boiled and let settle. The clear liquid is decanted and the precipitate, washed by decantation twice, using warm water. The liquid decanted off need not be filtered; it should be perfectly clear. Now add to the precipitate 10cc more HCl, which will dissolve it. Dilute to 200cc and again precipitate, and wash by decantation till free from HCl. Finally transfer to a filter, dry, ignite and weigh as Al_2O_3. The precipitate contains iron, phosphoric acid and titanic acid, but these bodies being only present in traces in ordinary slags, may be neglected.

The alumina precipitate as above obtained can be handled rapidly. Al_2O_3 is completely precipitated by NH_4HO without boiling in solutions containing a *large excess* of NH_4Cl, and when so precipitated settles easily.

Lunge, Jour. Soc. Chem. Inds., vol. IX, p. 111.

The above scheme for CaO fails when the slag contains much manganese. This will come down in part with the magnesia and partly separate with the iron and alumina in the 500cc flask.

In this case, after solution of the evaporated mass in HCl, dilute to 200cc, and add carbonate of sodium till a slight precipitate forms. Redissolve this with a drop or two of HCl, then add 3 grms. of sodium acetate and boil till the Al_2O_3 separates. Dilute to 500cc, mix and settle as before. Filter off 200cc. Add 5 grms. of sodium acetate then bromine water and determine the manganese as in iron ores. Treat the filtrate from the manganese for CaO and MgO as before.

Sulphur and *iron* are to be determined in slags as in iron ores.

The sulphur in slags is present principally as calcium sulphide. It may be determined approximately by adding 150cc of water to 0.5 grm. of the very finely pulverized slag and titrating with the standard iodine solution used for sulphur in iron.

Stir the mixture of slag and water well, add 3 or 4cc of starch solution, then run in the iodine till the blue color develops. Now add 15cc of conc. HCl, stir and add the iodine again until the color no longer disappears.

If 1cc of the iodine equals .0005 S, each cc taken will be equivalent to 0.1% sulphur in the slag.

Jour. An. and App. Chem., vol. VII, No. 5.

THE ANALYSIS OF FIRE CLAYS.

Determination of SiO_2, Al_2O_3, Fe_2O_3, CaO and MgO.—
Take 1 grm., mix it with 8 grms. of dry Na_2CO_3. Put the mixture in a platinum crucible, and heat over a Bunsen burner until the mass is shrunken together and caked. Now apply a blast lamp and rapidly fuse. Keep some minutes in quiet fusion. Cool suddenly by dipping the bottom of the crucible into water (this will usually cause the cake of material to come loose so that it can be easily detached from the crucible). Now put the cake, with the crucible, into a dish or caserole with water enough to cover the whole, and wash the cover of the crucible free from all attached particles. As soon as all is disintegrated, remove the crucible, cleaning it if necessary with a little HCl, which is then to be added to the body of the liquid. Now add an excess of HCl to the contents of the dish, which must be kept carefully covered to avoid loss by "spurting." Warm until everything is dissolved and effervescence has stopped. Wash off the cover and evaporate the liquid to dryness on a water bath, stirring occasionally to break up the jelly. When all odor of HCl has ceased, add a little water and again evaporate to dryness. Now add 30cc of dilute HCl (1 : 1) digest at a gentle heat, and then dilute to 150 or 200cc, filter and wash thoroughly, ignite and weigh repeatedly until the weight is constant. The residue is SiO_2 — its purity must be tested by volatilizing it by HFl and H_2SO_4, or by re-fusion with Na_2CO_3, and repeating the separation.

The filtrate should be diluted to 300cc, warmed and precipitated by a slight excess of NH_4HO. Warm until the precipitate has settled, leaving a perfectly clear liquid. Decant this off as far as possible without disturbing the precipitate (the use of a siphon for this purpose is often advantageous). Fill up to about 250cc with hot water and let settle again. Decant this in the same way. Repeat this until the liquid decanted off no longer reacts for HCl with $AgNO_3$.

Finally transfer the precipitate of $Al_2O_3 + Fe_2O_3$ ($+.TiO_2$ + traces of SiO_2) to a filter (do not wash on the filter) dry, transfer the precipitate to a crucible, carefully burn the paper *separate*, add the ash to the crucible, ignite strongly and weigh.

After weighing, brush the Al_2O_3, etc., out into a small beaker, cover it with a mixture of strong H_2SO_4 8 parts, water 3 parts, and digest on a hot plate for some time. All will dissolve but a slight residue of SiO_2. Dilute the liquid, then filter off and weigh this; it is to be deducted from the weight of the precipitate first obtained. In the liquid determine the *iron* volumetrically, as in iron ores, calculate it as Fe_2O_3 and deduct this, and the remainder will be Al_2O_3.

Collect all the washings from the Al_2O_3 except the first, separate from the first liquid decanted off, boil them down rapidly in a large porcelain dish to a small bulk and transfer them to a beaker. Now add a few drops of NH_4HO and filter from any trace of Al_2O_3 separating, use a small ashless filter. Add this filtrate to the first portion which was not boiled down, and the precipitate to the main portion of the Al_2O_3. Should the liquid settle badly as the washing proceeds, add 2 or 3 drops of NH_4HO, which will cause it to clear promptly.

Now concentrate the total filtrate from the Al_2O_3 to about 200cc on a water bath and determine the CaO and MgO exactly as in the analysis of a limestone.

Determination of the Alkalies by J. Lawrence Smith's Method.—

When silicates containing K_2O and Na_2O are heated with a mixture of $CaCO_3$ and NH_4Cl, $CaCl_2$ is first formed by double decomposition, and this then acts on the silicates forming alkaline chlorides and lime silicates. A red heat is necessary.

There is needed pure $CaCO_3$, free from K_2O and Na_2O. This can be prepared by dissolving marble in HCl, to saturation, adding a little slaked lime to make the liquid alkaline and precipitate Fe_2O_3, Al_2O_3 and P_2O_5, then diluting and heating the liquid and precipitating the $CaCO_3$ by $(NH_4)_2 CO_3$. This is washed till free from HCl and dried. Second, pure NH_4Cl. This must be powdered and must volatilize without residue at a low red heat.

Process.—Mix one grm. of clay with one grm. of NH_4Cl. Grind them together in a small porcelain mortar. Add eight grms. of $CaCO_3$ and mix thoroughly with the clay and $CaCO_3$. Take a large (30–50cc) platinum crucible, put a little pure $CaCO_3$ on the bottom, and then add the mixture. Clean out the mortar by grinding a little more $CaCO_3$ in it. Add this on top of the mixture as a cover.

Now cover the crucible and heat carefully, gently at first, but gradually to full redness for an hour. Cool and transfer the sintered mass to a beaker. Wash the crucible and cover with hot water and add the washings. Digest the whole until the mass slakes down to a fine powder. Now filter and wash with hot water until the filtrate is free from Cl.

To the filtrate add NH_4HO and $(NH_4)_2 CO_3$ in excess. The calcium separates as carbonate, which on warming becomes granular and easily filtered. Filter and wash with water containing a very little NH_4HO.

Evaporate the filtrate to a few cc's in the beaker, then transfer it to a small porcelain dish and finally bring it to dryness.

Now ignite it carefully at a heat not exceeding a barely visible red, until all NH_4Cl is expelled and no more fumes form. Cool and add a little water and a few drops of $(NH_4)_2 CO_3$, and filter from any residue. Add two or three drops of HCl to the filtrate. Again evaporate to dryness in a weighed porcelain dish. Dry, ignite carefully as before and weigh as KCl + NaCl. The chlorides must be *white* and dissolve without residue in water.

To the water solution add an excess of platinum chloride, and evaporate carefully to nearly dryness. Add 20cc of alcohol (80 per cent.) and let stand till the Na salts dissolve. Filter on to a weighed filter. Wash the K_2PtCl_6 with 80 per cent. alcohol. Dry and weigh. Calculate the K_2O from the weight of this and the Na_2O from the remainder of the mixed chlorides after deducting the KCl calculated from the K_2O.

The strength of the alcohol is important. The K_2PtCl_6 is practically insoluble in 80 per cent. alcohol, but the Na_2PtCl_6 will dissolve in it. Time must be given to secure complete solution of this latter salt.

Titanium may be determined in fire clay as in the insoluble residue from titaniferous iron ores.

Clay contains silica as fine sand or quartz, also silica in combination with alumina. It is sometimes desirable to find the amounts of these separately. The quartz is insoluble in potash solution, while the SiO_2 left after evaporation of solutions of silica in HCl is soluble, also that from the decomposition of silicates by H_2SO_4.

Process for Determination of Free and Combined Silica.—Treat 1 grm. of the clay with 10 or 15cc of conc. H_2SO_4. Heat to near the boiling point of the acid and digest for 12 hours. Cool, dilute, filter, wash, ignite to constant weight.

The residue consists of SiO_2 as sand, SiO_2 from the decomposition of the silicates of alumina and some insoluble silicates. Transfer to an agate mortar, grind fine, brush on to a watch glass and weigh again.

Heat 50cc of a 15% solution of KHO to boiling in a platinum dish. Add the above weighed residue and boil five minutes. Again filter, wash, ignite and weigh. The silica which had been separated from the alumina will have dissolved the residue being sand and silicates. Deduct this weight (calculated to the whole residue) from the original weight of residue and the difference will be the *combined silica*. This in turn deducted from the total silica, regularly determined, gives the silica as sand and undecomposable silicates.

The above scheme for fire clay analysis avoids the use of hydrofluoric acid. Where that is available the decomposition of the clay by its use both for alumina and alkalies is more rapid.

See Blair, Chem. An. Iron and Steel, Second Edition.

THE DETERMINATION OF COPPER IN ORES.

Most copper ores are dissolved by digestion with conc. HNO_3. Sulphur may separate, but will be practically free from copper. Some few bodies (as slags) may require to be fused with sodium carbonate and nitrate.

The most generally reliable method for the determination of copper is by electro-deposition. The conditions most favorable are: a sufficiently dilute solution; not too much free acid, and not too strong a current, failure in these will cause the copper to be spongy, dark colored and difficult to wash.

Hydrochloric acid must be absent; also much nitric acid. Presence of much arsenic, antimony or silver will cause the copper to be impure. Bismuth is especially hurtful, very small percentages causing the results to be high. Should these elements be present in the ores they must be removed before precipitating the copper. They are rarely present in ordinary ores in quantity sufficient to influence the results.

Process for Copper Ores.—Take an amount of ore which shall not contain more than 0.200 grms. of Cu. It must be very finely pulverized. Put it in a small beaker, add 5cc HNO_3, 5cc HCl and 5cc H_2SO_4. Cover with a watch glass and boil until the ore is decomposed, all the HCl and HNO_3 are expelled and white fumes of H_2SO_4 begin to appear. Cool, dilute to 50cc and boil, filter and wash. The residue should be light colored and must be tested for copper with the blow-pipe.

Put the solution in a weighed platinum dish of 100cc capacity. Connect the dish with the zinc side of a battery of two gravity cells or its equivalent, and introduce into the solution a platinum plate connected with the copper end of the battery. The deposition begins at once and is complete in seven or eight hours. When all trace of blue color is gone from the solution, take out a few drops with a pipette and test with H_2S water. If no tinge of brown is produced the Cu is all down. Empty the dish, wash it carefully first with distilled water, then with strong alcohol. Dry the alcohol off carefully, avoiding a temperature much above 100°C and weigh the dish plus the copper.

The alcohol may be lighted and "burned off" without danger, and the heat thus developed will dry the dish. The copper must be red, coherent and metallic. A slight brownish discoloration will not be of perceptible influence on the results. See

Peter's "American Methods of Copper Smelting," p. 25; also W. Lee Brown, Assaying, p. 333, et seq.

THE IODINE METHOD FOR COPPER.

This is much more rapid than the electrolytic method, and when carefully conducted gives very accurate results. It depends on the fact that when a cupric salt is treated with potassium iodide cuprous iodide is formed and iodine liberated. The iodine is then estimated volumetrically with sodium hyposulphite. The conditions are a slightly acid solution and absence of considerable amounts of sodium acetate. Iron must be absent and also any considerable amount of antimony or bismuth.

Preparation of the "Hypo" Solution.—Dissolve 39.18 grms. crystallized sodium hyposulphite in water and dilute to one liter. 1cc equals 0.01 grms. of copper.

The reactions from which this is calculated are as follows: $2\,CuSO_4 + 4\,KI = Cu_2I_2 + 2\,I + 2\,K_2SO_4$ and $2\,Na_2S_2O_3 + 2\,I = 2\,NaI + Na_2S_4O_6$. The crystallized hyposulphite contain five molecules of water. ($Na_2S_2O_3, 5\,H_2O$.)

Standardize the solution against pure copper or pure copper sulphate. Dissolve the copper in 5cc of dilute nitric acid (1 to 1), boil off all nitrous fumes (this is essential or they would liberate iodine) dilute with an equal bulk of water and add a solution of NaHO cautiously until a permanent precipitate is just produced. Now add 1cc of acetic acid, which must give a clear solution; *if the liquid is warm cool it*. Now add 3 grms. of pure potassium iodide. When it is dissolved dilute to 100cc. Now run the hypo solution from a burette until the brown color due to the liberated iodine is nearly discharged. Then add 2 or 3cc of starch solution and continue to add the hypo until the blue color is gone and *does not return on standing four minutes*. The number of cc of hypo used will be equivalent to the copper taken.

Process for Ores.—Take 2 grms. ore (very finely powdered). Heat in a small covered beaker or caserole with 20cc of conc. HNO_3. When violent action has ceased boil down to dryness. Take up with 30cc conc. HCl, digest and dilute to 200cc without filtering. Warm and pass a rapid current of H_2S through the solution till the Cu is all down as sulphide. Filter and wash with water containing a little H_2S. Wash the precipitate back into the beaker or dish. If more than a trace remains on the filter, burn it and add the ash to the rest. Now add 15cc of conc. HNO_3 and boil almost to dryness. Add 20cc of water and boil till all nitrous fumes are expelled. Filter from the gangue and sulphur. Neutralize the solution with NaHO solution as before. Cool it and add 5 grms. of KI, dilute to 100cc and titrate with the hypo solution. The sulphur and gangue on the filter should be burned off in a porcelain crucible and the residue examined for copper.

See "A Text Book of Assaying," C. and J. J. Beringer, p. 150, et. seq.

THE ASSAY OF ORES FOR ZINC.

Zinc usually occurs in ores as sulphide, oxide, carbonate or hydrosilicate.

All of these are decomposed by boiling with acids, the zinc passing into solution. When sulphur is present it can be oxidized by HNO_3 and $KClO_3$.

Acid solutions of zinc in HCl are completely precipitated by potassium ferrocyanide. Iron, copper, cadmium and manganese are also precipitated in the same way, and if present must be removed from the solution before the zinc is determined.

An excess of ferrocyanide in the solution can be recognized by the brown precipitate it gives with a solution of uranic nitrate or acetate (uranic ferrocyanide). This reaction is less sharp in acid solution, but by using a concentrated solution of the uranium salt, and only a little of the solution to be tested, is still sufficiently delicate.

The HCl solution must be free from Cl or oxide of chlorine, as these decompose the ferrocyanide and liberate iron salts.

PROCESS OF VON SCHULZ AND LOW.

Solution of the Ore and Separation of the Fe, Mn and Cu.—Weigh one grm. into a 4-inch caserole. Add 2 or 3cc of conc. HNO_3, then cautiously 25cc of HNO_3, previously saturated with $KClO_3$ by shaking up with crystals of the salt. (Keep this solution in an open bottle). When the violent action is over, cover the caserole and boil rapidly to dryness. Do not bake the residue. Now cool and add seven grms. of NH_4Cl, 25cc of hot water and 15cc of strong NH_4HO. Boil the liquid one minute and then rub the dish with a rubber tipped rod to loosen and disintegrate all the insoluble matter. Filter and wash several times with a boiling hot 1 per cent. solution of NH_4Cl. If the filtrate is *blue* it contains Cu.

Add to the filtrate 25cc of conc. HCl and dilute to 200cc. If Cu is present add forty grms. of "granulated lead" and stir until the liquid is colorless.

The zinc salts are soluble in NH_4HO, while the Fe_2O_3, Al_2O_3, etc., are precipitated. This separation is quite satisfactory for ores containing moderate percentages. When much zinc is present or much iron, the residue may retain enough to affect the results 1 or 2 per cent. In this case it must be dissolved in 10 or 15cc of HCl, and reprecipitated by NH_4HO.

This second filtrate will contain Mn if present in the ore. To remove it add 5 or 10cc of hydrogen peroxide (H_2O_2) and filter from the MnO_2. Add this filtrate to that from the main quantity.

The HNO_3 and $KClO_3$ separate all the Mn in the first case as MnO_2, but this dissolves in the HCl again.

It is essential that all the $KClO_3$ be decomposed and the Cl driven off in the evaporation, as if any gets into the final solution for titration it will cause the solution to become blue and use up some ferrocyanide.

This may be prevented by somewhat more prolonged heating of the dry residue until any $KClO_3$ is decomposed, or by adding a little *sodium sulphite* to the solution before titration.

The granulated lead precipitates the Cu as metal. The lead in solution does not interfere with the subsequent titration.

Volumetric Determination of the Zinc — Preparation of the Ferrocyanide Solution.—Dissolve forty-four grms. of pure $K_4Fe(CN)_6 \cdot 3 H_2O$ in water and dilute to one litre. 1cc of this will precipitate approximately 0.01 grm. of Zn.

Standardize the solution against pure zinc. Dissolve 0.2 grms. in 10cc of HCl, add seven grms. of NH_4Cl, dilute to 100cc and heat to boiling. Now run in the ferrocyanide solution until a drop of the liquid shows a brown tinge when tested on a white plate with a drop of a strong solution of uranic nitrate after standing two or three minutes. To save time make several such tests $\frac{1}{10}$cc apart, and take the reading of the one showing color in three minutes. Now make a similar test upon the HCl + NH_4Cl without the zinc. Deduct the amount of this "blank" from the other, and the difference gives the amount of the solution, equivalent to 0.2 grms. zinc.

Determination of the Zinc.—Titrate the strongly acid solution of the ore exactly as above. The liquid must be boiling hot.

To prevent "running over" it is a good plan to take out one-third of the liquid. Titrate the remainder roughly, and then add the one-third and finish carefully.

Cadmium when present counts as zinc in this process.

See Jour. An. and App. Chem., vol VI, p. 491.
Beringer, Assaying, p. 217.

THE ANALYSIS OF ALLOYS OF LEAD, ANTIMONY, TIN AND COPPER. (BEARING METAL.)

Traces of As, Fe and Zn are often present. The complete analysis of these alloys is beyond the scope of these notes, but the following method for the determination of the four principal metals will be found satisfactory if carefully conducted.

The most troublesome operations are the separation of the tin from the lead and from the antimony.

See Fresenius Quant. Anal., ₴ 164, also ₴ 165, ₴ 126 and ₴ 125.

Process.—Weigh 0.5 grms. of the fine shavings into a 150cc beaker. Add two grms. of solid tartaric acid (powdered), and then 15cc of HNO_3, 1.2 sp. gr. Cover and warm until everything is dissolved, wash off and remove the cover and evaporate carefully to a pasty mass. Now add 50cc of water and warm until all the lead nitrate has dissolved. The tin and antimony, in part, are left as a fine white powder. Now drop in a conc. solution of KHO until the precipitate formed is dissolved in the excess. A cloudiness may remain, but almost all will go into solution. Add 10cc of "yellow sulphide of sodium or potassium,"[1] set on a water bath and digest at a temperature considerably short of boiling for three or four hours, stirring occasionally and keeping the beaker covered. Now decant the clear liquid through a filter, and wash once by decantation. Avoid getting the precipitate on the filter, but decant close each time. To the residue add 10cc more of the sulphide solution, but no water and digest again for two hours. Then add 50cc of water and warm, let settle, decant, transfer and wash the precipitate with H_2S water.

The Precipitate contains PbS and CuS. Dry the filter and precipitate; detach the latter as completely as possible and burn the filter in a small porcelain crucible using a very low heat and merely driving off volatile matter, not attempting to completely burn the carbon. Now add the burned filter to the rest of the PbS in a small casserole. Cover and add a few drops of conc. HNO_3 to moisten the PbS; then add 5cc of *fuming* HNO_3 (1.5 sp. gr.). Warm for some time and, when all sulphur has disappeared, add 5cc of H_2SO_4 (1 : 3) and evaporate till all HNO_3 is expelled. Now add 25cc of water; stir well, let settle, filter and wash with water containing about 1% H_2SO_4, and finally wash with alcohol. The precipitate is $PbSO_4$. When it is dry, detach it carefully from the filter and put in a small weighed porcelain crucible. Burn the filter paper carefully on the inverted lid of the crucible and

---

1. *Sulphide of Sodium Solution.*—This is made by saturating a 20 per cent. solution of sodium hydrate with H_2S gas. Then filter the solution and add to each 100cc about 100 millegrammes of flowers of sulphur. This will dissolve and make the liquid yellow. This solution must be kept in filled, closely stopped bottles and used fresh.

add to the ash one or two drops of conc. HNO_3 and one drop of H_2SO_4, evaporate off the acids carefully, and finally dry and ignite the crucible and cover at a low red heat. Now weigh all together and determine the $PbSO_4$, from which calculate the Pb.

The filtrate from the $PbSO_4$ contains the Cu. This may be precipitated by the battery, or it may be thrown down by a current of H_2S, filtered out, washed with H_2S water, dried, ignited and weighed as $CuO + Cu_2S$. (The fact that the weighed precipitate is partially converted to oxide by the air is unimportant, as the per cent. of Cu in CuO and Cu_2S is the same).

The Solution from the PbS and CuS contains the Sb and Sn (and As) as sulphides. Dilute the liquid to about 250cc. Add HCl carefully until the solution distinctly reddens litmus, but avoid much excess, set on a warm plate and heat *gently* until the odor of H_2S has nearly gone.

Filter and wash once by decantation, then run the precipitate on the filter. The filtrate will grow milky in time, but no precipitate should separate from it.

The precipitate, consisting of Sb_2S_3, SnS_2 and free sulphur, is now washed back into the beaker, using a wash bottle with a small jet, so as to accomplish this with but little water. With care the amount of precipitate left on the filter paper will be but trifling. Now dissolve this off *through the filter* into the beaker with a few drops of a dilute solution of NaHO; the liquid in the beaker need not be more than 75cc. Add to the contents (water and sulphides) 10 to 15 grms. of solid NaHO; when this is dissolved, add carefully about 3cc of *bromine* (not "bromine water.") Now digest on a water bath keeping the beaker covered, until the sulphur is oxidized or collected in a granular form and the antimony has separated as a white crystalline precipitate of sodium metantimoniate.

Test the liquid by adding to a drop of it a drop or two of HCl, if it gives off *bromine vapor* enough Br. has been added; if not, add a little more. Finally boil the liquid a few minutes. Now cool and add one-third the volume of alcohol. Let stand some hours, filter and wash with water containing one-third its volume of alcohol and a little Na_2CO_3. The filtrate contains all the SnO_2 as sodium stannate, and the precipitate contains the antimony and often some free sulphur.

Dilute the *filtrate* to 200cc, add HCl until it is distinctly acid, now warm till any Br. disappears, and then add 200cc of H_2S water, and pass H_2S gas till saturated. Let stand till the precipitate settles, filter and wash the SnS_2 thoroughly. If the precipitate tends to run through the filter, add ammonium acetate to the wash water. Put the filter and precipitate in a weighed porcelain crucible, dry and burn off the paper very carefully, finally ignite and weigh as SnO_2. Add a little solid ammonium carbonate to the crucible; heat, ignite intensely and weigh again. Any loss is due to sulphuric acid held by the SnO_2.

Notes on Metallurgical Analysis.

The Antimony Precipitate is washed back into the beaker and dissolved in the least possible amount of dilute HCl, containing a little tartaric acid. Wash the filter with the same, and finally filter the solution from any residual sulphur. (Which will be free from antimony.)

Now dilute to about 250cc, heat to nearly boiling and pass a rapid current of H_2S till the Sb_2S_3 is all precipitated.

Weigh a 7 c. m. filter paper, as in the phosphorus determination by the yellow precipitate method. Let the Sb_2S_3 precipitate settle, decant through the filter. Transfer the precipitate, wash well and dry carefully at 100°C and weigh filter, plus precipitate. This gives the total weight of the precipitate, which always contains free sulphur. Now detach as much as possible from the filter paper, transfer it to a weighed porcelain boat and weigh the boat and contents.

Take a piece of combustion tubing large enough to receive the boat and its contents and ten or twelve inches long. Connect one end with a flask for generating CO_2 and containing fragments of marble. Close the other end of the combustion tube with a cork containing a small exit tube. (There must be a U-tube containing $CaCl_2$ between the flask and the combustion tube to remove moisture from the gas.)

Now put the boat and its contents about the middle of the tube, pour a little dilute HNO_3 into the CO_2 flask and let the current of gas pass till all air is expelled (four or five minutes). Now carefully heat the tube around the boat. The free sulphur volatilizes and the Sb_2S_3 turns black and metallic looking. When this change is complete and no more sulphur vapors pass off, cool the apparatus, withdraw the boat and weigh it. The weight of the contents gives the pure Sb_2S_3 in the amount taken.

From this calculate the Sb_2S_3 in the total precipitate as weighed originally, and from this again the Sb in the sample.

Notes on the Above Scheme.—The use of the tartaric acid is to assist the solution of antimony, which is but slowly oxidized by HNO_3 alone. The separation of tin by the solubility of the bisulphide in sodium sulphide is complete, provided there is repeated and prolonged digestion with large excess. Sulphide of sodium is used instead of sulphide of ammonium as the latter dissolves copper sulphide. In the treatment of the PbS the use of the fuming nitric acid causes the complete oxidation of the sulphur. With ordinary conc. nitric acid, which contains about 30% of water, there will more or less sulphur separate which fuses into globules. When the $PbSO_4$ is ignited this melts and causes reduction to sulphide, and loss of weight.

The fuming acid can be most easily prepared by mixing in a large retort one part of ordinary HNO_3 (C. P.) with two parts of conc. H_2SO_4 and distilling off the nitric acid into a cooled dry receiver. The neck of the retort should project well into the receiver.

Arsenic, if present, is found with the stannic sulphide. It is driven off on heating the SnO_2.

The foregoing method for the separation of tin and antimony depends upon the oxidation of the Sb_2S_3 by the sodium hypobromite to sulphuric acid and sodium metantimoniate; the usual method of effecting this is by fusion with sodium hydrate and nitrate, after preliminary oxidation by HNO_3. The wet method above given is much less troublesome and equally complete. For the fusion method see Fres. Quant. loc. cit.

Clark's method of separating the sulphides by oxalic acid is also convenient and gives excellent results. The process is as follows:

Wash the sulphides into a beaker as before, then treat them with $KClO_3$ and HCl. Evaporate to dryness and add tartaric acid, then HCl and water. Filter from any residue of sulphur. Dilute and add oxalic acid in large excess then precipitate by H_2S. The Sb_2S_3 comes down, the Sn stays in solution. The precipitation of the tin in the filtrate is a matter of some difficulty, but can be accomplished if the excess of acid be neutralized and the solution saturated with H_2S, made ammoniacal, then acidified with acetic acid.

See Am. Jour. Sci., vol. XLIX, p. 154, and Am. Chem. Jour., vol. I, p. 244.

THE EXAMINATION OF WATER FOR BOILER SUPPLY.

The determination of the "scale-forming ingredients" and in the case of water contaminated with mine drainage, the "acidity" of the water is all that is necessary.

Outline Process for the Analysis.—First, evaporate 100cc of the clear (if necessary filtered) water to dryness in a weighed platinum dish and dry at 100° to constant weight. This gives the "total solids." Second, test the water for chlorine. If any considerable amount is found determine it volumetrically with a standard solution of $AgNO_3$, adding a little neutral potassium chromate to the water to serve as an indicator. The slightest excess of $AgNO_3$ gives the reddish color due to silver chromate. (See Fresnius Quant., ₴141.) Third, acidulate one litre of the water with a little HCl and evaporate it to about 175cc. Transfer it to a 200cc flask and dilute to the mark. Take of this 100cc, evaporate to dryness and determine the silica, iron and alumina, the CaO and the MgO exactly as in a limestone. Of course reducing the volumes and amounts of reagents to correspond to the smaller quantity taken. Or the evaporation to dryness may be omitted and the silica will then come down with the Fe_2O_3 and may be so included. Take 50cc and determine the SO_3 as $BaSO_4$ by precipitation with $BaCl_2$. (See determination of sulphur).

Now calculate the results as follows: Combine the sulphuric acid first with the calcium. All calcium left over is to be estimated as carbonate. The magnesium is to be combined first with the chlorine; next with any sulphuric acid left from the calcium and lastly the residue estimated as carbonate.

Should the water contain alkalies in any amount, this will be indicated if the weight of the "total solids" exceeds by any considerable amount the sum of the sulphates, carbonates and chlorides of calcium and magnesium, the oxides of iron and alumina, and the silica. They may be determined in the dry residue as follows:

Boil the residue with a few cc of water, filter and wash. Treat the solution first with an excess of BaH_2O_2 solution and filter, then with ammonium carbonate. Filter the solution from the precipitate of $BaCO_3$. Evaporate the filtrate and expel the ammonium salts by ignition. Determine the alkalies in the residue as in the similar residue in the analysis of fire clays.

In computing results in this case, first combine the alkalies with the sulphuric acid and chlorine, proceeding with the remainder as before.

APPENDIX.

TABLE OF ATOMIC WEIGHTS.

The following table comprises the atomic weights of all the elements of usual occurrence or of special interest to the metallurgist. It is taken from one published by the U. S. Dep't. of Agriculture, and revised by F. W. Clark, Chief Chemist of the U. S. Geological Survey, and represents the most trustworthy results up to 1890.

Name.	Symbol.	Atomic weight.	Name.	Symbol.	Atomic weight.
Aluminum	Al	27	Manganese	Mn	55
Antimony	Sb	120	Mercury	Hg	200
Arsenic	As	75	Molybdenum	Mo	96
Barium	Ba	137	Nickel	Ni	58.7
Bismuth	Bi	208.9	Nitrogen	N	14.03
Boron	B	11	Oxygen	O	16
Bromine	Br	79.95	Palladium	Pd	106.6
Cadmium	Cd	112	Phosphorus	P	31
Calcium	Ca	40	Platinum	Pt	195
Carbon	C	12	Potassium	K	39.11
Chlorine	Cl	35.45	Silicon	Si	28.4
Chromium	Cr	52.1	Silver	Ag	107.92
Cobalt	Co	59	Sodium	Na	23.05
Copper	Cu	63.6	Strontium	Sr	87.6
Fluorin	F	19	Sulphur	S	32.06
Gold	Au	197.3	Tin	Sn	119
Hydrogen	H	1.007	Titanium	Ti	48
Iodine	I	126.85	Tungsten	W	184
Iron	Fe	56	Uranium	U	239.6
Lead	Pb	206.95	Vanadium	V	51.4
Lithium	Li	7.02	Zinc	Zn	65.3
Magnesium	Mg	24.3			

Appendix. 99

TABLE OF FACTOR WEIGHTS.

The following table shows the amounts of material to be taken that each milligramme of the precipitate weighed may equal some definite percentage of the substance sought:

Substance to be determined.	Precipitate to be weighed.	Amount of the substance to be determined in the precipitate weighed.	Factor weights to be taken.	Per cent. of the substance to be determined represented by each milligramme of precipitate.
S	$BaSO_4$	0.1373	1.373	0.01
SO_3	$BaSO_4$	0.3433	1.716	0.02
Si	SiO_2	0.4667	1.167	0.04
P	$Mg_2P_2O_7$	0.2790	2.790	0.01
P_2O_5	$Mg_2P_2O_7$	0.6396	2.198	0.03
Mn	Mn_3O_4	0.7200	1.440	0.05
Mn	$Mn_2P_2O_7$	0.3873	0.387	0.10
C	CO_2	0.2727	2.727	0.01

APPLYING THE "CORRECTION FACTORS" OF VOLUMETRIC SOLUTIONS TO THE AMOUNTS OF SUBSTANCE TAKEN.

This method of avoiding calculation in obtaining the results of volumetric analyses is frequently very convenient.

It has been indicated in the notes on the iodine method for sulphur.

In general where a solution is made up so that each cubic centimeter shall represent a definite per cent. of some constituent when a certain weight of substance is taken for the analysis; if on standardizing the solution each c c is found to be in fact equivalent to some fraction of that value, the usual plan is to multiply all results by that fraction to get the true results.

If, however, the assumed amount of substance be multiplied by that same fraction (or factor), and the resulting amount actually used in the analysis, the number of cubic centimeters taken will give the true percentages at once.

Thus in the Emmerton phosphorus method when 1 c c of the permanganate is calculated to be equal to 0.002 per cent. of P if five grms. of iron are taken, suppose that on standardizing 10.3 c c of the solution were found equivalent to what 10 c c should be.

Then 1 c c would be equivalent to $\frac{10.}{10.3}$ of .002 per cent. on five grms. taken, $\frac{10}{10.3}$ being the "factor." In this case by "weighing out" 5 $\times \frac{10.0}{10.3}$ grms. (4.854), the number of c cs used, multiplied by .002, will give the percentage.

SOME ADDITIONAL NOTES AND METHODS.

1. SAMPLING SPIEGEL IRON AND WHITE CAST IRON.

These and similar materials which are too hard to drill must be broken into small fragments with a sledge hammer and several pieces pulverized in a steel mortar. A very efficient mortar for this purpose can be made by boring an inch hole two inches deep into a block of tool steel about three inches square and four inches high. Fit this with a steel "rammer" cut from a round bar and about three inches longer than the hole. It must be only slightly smaller than the hole in the block. Both block and rammer must be well hardened. By dropping a fragment of metal into the hole, inserting the rammer and pounding it vigorously with a heavy hammer the hardest material is soon reduced to a fine sand.

2. THE DETERMINATION OF IRON IN ORES BY PERMANGANATE.

The following very rapid method is used on Lake Superior ores:
(See Eng. and Min. Jour., Vol. LVII, No. 15.)

Take about 0.5 grm. of ore, add $2\frac{1}{2}$ c c of $SnCl_2$ solution and 10 to 15 c c of HCl (1:1). Boil gently until the iron is all dissolved. This is easily seen owing to the light color of the solution, the $SnCl_2$ reducing the iron to the ferrous form. Now, if necessary, drop in more tin solution to complete the reduction and then add 5 c c of a saturated solution of $HgCl_2$ to remove the excess of $SnCl_2$. Dilute the solution to about 250 c c, add 5 or 10 c c of "titrating solution," then add standard permanganate solution until the last drop gives a persistent pink color.

The "titrating solution" is made by dissolving 160 grms. of manganous sulphate in water, diluting to 1750 c c and adding 330 c c of phosphoric acid and 320 c c of sulphuric acid.

The standard permanganate solution is calculated so that each cubic centimeter shall represent 2 per cent. of iron when 0.5 grm. of ore is taken. Its exact value is then determined by testing against a known ore or pure iron, and then the amount of ore taken in the analysis so varied that each c c of the actual solution still represents 2 per cent. The addition of the $SnCl_2$ during solution reduces the time required. The sulphate of manganese in the "titrating solution" prevents the reduction of permanganate by the HCl present. The process would have to be used with caution, as the presence of organic matter or other reducing agents in the ore would render the results inaccurate.

3. ON DIFFICULTY IN FILTERING SOLUTIONS OF PIG IRON AND STEEL IN PHOSPHORUS DETERMINATIONS.

When pig iron or steel high in silicon is dissolved in HNO_3 and after evaporation and "baking" the residue dissolved in HCl, the solution obtained will be found very slow in filtering owing to the presence of SiO_2 in a peculiarly gelatinous form.

The SiO_2 left after the evaporation of an HCl solution is much more granular and easily filtered off. Therefore, in all cases where silicon is present to any extent the HCl solution of the baked residue must be evaporated to dryness again and then redissolved in HCl. This second evaporation takes but little time, and is essential when using a process like that of Emmerton or Wood on cast iron, if a long and tedious filtration is to be avoided.

It is stated that the addition of a few drops of HF or a little NH_4F to the first HCl solution will also cause it to filter more rapidly and render the second evaporation unnecessary.

4. THE PURIFICATION OF BARIUM SULPHATE.

When a precipitate of $BaSO_4$ is colored red by iron, as is sometimes the case in sulphur determinations when the solution was too hot or contained too little free acid, it can be easily purified by solution in concentrated H_2SO_4. Add to the precipitate in the crucible 2 or 3 c c of concentrated H_2SO_4 and heat till solution is complete. Cool and pour into 100 c c of cold water, washing out the crucible. The $BaSO_4$ separates immediately and can be filtered off, washed a little, ignited and weighed. The results are, however, liable to be too low, not from the failure to recover all the $BaSO_4$, but because the red precipitate may contain basic iron sulphate.

5. ON THE PRESENCE OF NITRITES IN CAUSTIC ALKALIES AS A SOURCE OF ERROR IN SULPHUR AND CARBON DETERMINATIONS.

Commercial caustic potassa sometimes contains *nitrites*. The effect of these on the volumetric sulphur process is to liberate iodine from the KI in the solution, and so vitiate the results. Their presence is not shown by the ordinary "blank" test with starch and iodine solution.

To test the potash solution add HCl until acid, then starch paste and a little pure KI; if a blue color develops, a nitrite or some similar compound is present and the solution cannot be used. These and other oxygen absorbents like FeO, when present in caustic potassa and soda, also cause trouble in the "carbon train," giving rise to constant gain in weight in the potash bulbs during aspirating. The difficulty may be overcome by adding permanganate of potassa to the boiling alkali solution drop by drop until a faint, persistent green color is produced. Let the liquid cool and settle and decant the clear solution.

ERRATA.

Page 14, *line* 3, *for* CaH_2O_2, *read* MgH_2O_2.
Page 14, *line* 30, *for* CaC_4O_4, *read* CaC_2O_4.
Page 15, *line* 13, *for* Fe_2Cl_4, *read* Fe_2Cl_6.
Page 16, *line* 33, *for* basis, *read* basic.
Page 20, *line* 5, *for* 130 Cy, *read* 130°C.
Page 20, *line* 5, *for* 12 MoO_3PO_4 $(NH_4)_3$, *read* 12 MoO_3, PO_4 $(NH_4)_3$.
Page 32, *line* 26, *for* 20 : n = .0001 : x, *read* n : 20 = .0001 : x.
Page 33, *line* 19, *for* HNO_4, *read* HNO_3.
Page 35, *line* 4 from bottom, *for* one-fourth, *read* one.
Page 40, *line* 17, *for* now dilute to 150cc and add NH_4HO till a slight permanent precipitate is formed, then a few drops of acetic acid and boil. *Read* now dilute to 150 c c, add NH_4HO till just alkaline, then acetic acid drop by drop till just acid, and boil.
Page 41, *line* 6, *for* drillings, *read* powdered metal.
Page 41, *line* 6, *for* dissolve in 10 c c HNO_3 1.2 sp. gr. evaporate—*read* dissolve in 10 c c HNO_3 1.2 sp. gr. and 5 c c HCl evaporate—
Page 52, *line* 25, *for* add 5 c c conc. HCl, *read* add 5 to 10 c c conc. HCl.
Page 52, *line* 31, *for* now heat, *read* now warm to about 60°.
Page 57, *line* 14, *for* $Na_2S_2O_3$ 5 H_2O, *read* 2 $Na_2S_2O_3$ 5 H_2O.
Page 72 to references on color.TiO_2, add Dunnington Jour. Am. Chem. Soc., XIII, No. 7.
Page 84, *line* 6, *for* warmed, but *not* boiled, *read* boiled for 2 or 3 minutes.
Page 85, *line* 27, *for* warmed, *read* boiled.
Page 87, *line* 12, *for* now filter and wash with hot-water until the filtrate is free from Cl, *read* now filter and wash with hot water until the filtrate amounts to about 250 c c. This will extract all the alkalies.